T0214756

SpringerBriefs in Molecular Science

Biobased Polymers

Series editor

Patrick Navard, CNRS/Mines ParisTech, Sophia Antipolis, France

Published under the auspices of EPNOE*Springerbriefs in Biobased polymers covers all aspects of biobased polymer science, from the basis of this field starting from the living species in which they are synthetized (such as genetics, agronomy, plant biology) to the many applications they are used in (such as food, feed, engineering, construction, health, …) through to isolation and characterization, biosynthesis, biodegradation, chemical modifications, physical, chemical, mechanical and structural characterizations or biomimetic applications. All biobased polymers in all application sectors are welcome, either those produced in living species (like polysaccharides, proteins, lignin, …) or those that are rebuilt by chemists as in the case of many bioplastics.

Under the editorship of Patrick Navard and a panel of experts, the series will include contributions from many of the world's most authoritative biobased polymer scientists and professionals. Readers will gain an understanding of how given biobased polymers are made and what they can be used for. They will also be able to widen their knowledge and find new opportunities due to the multidisciplinary contributions.

This series is aimed at advanced undergraduates, academic and industrial researchers and professionals studying or using biobased polymers. Each brief will bear a general introduction enabling any reader to understand its topic.

*EPNOE The European Polysaccharide Network of Excellence (www.epnoe.eu) is a research and education network connecting academic, research institutions and companies focusing on polysaccharides and polysaccharide-related research and business.

More information about this series at http://www.springer.com/series/15056

Majda Sfiligoj Smole · Silvo Hribernik ·
Manja Kurečič · Andreja Urbanek Krajnc ·
Tatjana Kreže · Karin Stana Kleinschek

Surface Properties of Non-conventional Cellulose Fibres

 Springer

Majda Sfiligoj Smole
Faculty of Mechanical Engineering
University of Maribor
Maribor, Slovenia

Manja Kurečič
Faculty of Mechanical Engineering
University of Maribor
Maribor, Slovenia

Tatjana Kreže
Faculty of Mechanical Engineering
University of Maribor
Maribor, Slovenia

Silvo Hribernik
Faculty of Mechanical Engineering
University of Maribor
Maribor, Slovenia

Andreja Urbanek Krajnc
Faculty of Agriculture and Life Science
University of Maribor
Hoče, Slovenia

Karin Stana Kleinschek
Faculty of Mechanical Engineering
University of Maribor
Maribor, Slovenia

ISSN 2191-5407   ISSN 2191-5415   (electronic)
SpringerBriefs in Molecular Science
ISSN 2510-3407   ISSN 2510-3415   (electronic)
Biobased Polymers
ISBN 978-3-030-10406-1   ISBN 978-3-030-10407-8   (eBook)
https://doi.org/10.1007/978-3-030-10407-8

Library of Congress Control Number: 2018965435

This Springer imprint is published by the registered company Springer Nature Switzerland AG
The registered company address is: Gewerbestrasse 11, 6330 Cham, Switzerland

Contents

About the Authors

Majda Sfiligoj Smole is a Professor in the field of material science. She obtained her doctoral degree at the Karl Franzens University in Graz, Austria, in the field of physical chemistry. At the Faculty of Mechanical Engineering, University of Maribor, she lectured in natural and man-made fibres, structure of materials, sustainable materials and nanotechnology. From 2012 to 2016, she was the Head of the Chair for Textile Materials and Design at the Faculty of Mechanical Engineering, University of Maribor. Her main research topic is on study of fibre structure—properties relationships and fibres' modification for their functionalization. Recently, her research was focused mainly on nanotechnological modification of textiles and sustainable textile materials. She has published more than 160 scientific articles, conference papers, monographs and book chapters and is a co-author of three patents. In addition, she is a Member of the European network of excellence for polysaccharides EPNOE and European Textile Technology Platform.

Silvo Hribernik is a Scientific Associate at the Faculty of Mechanical Engineering and Faculty of Electrical Engineering and Computer Science, University of Maribor, working on several national and international basic and applied projects. He finished his graduate study in 2005 in the field of eco-textile engineering and doctoral study in 2010 in the field of textile technology, both conducted at the Faculty of Mechanical Engineering in Maribor, Slovenia. During this time, he also performed research work as a Visiting Scientist at Institute of Chemistry in Ljubljana and at the Fraunhofer Institute in Potsdam, Germany. He has so far published several scientific papers in JCR journals, has co-authored chapters in monographs, as well as participated and presented at a number of international conferences. Main area of his research is the development of fibre-based functional materials with the implementation of nanotechnology and is comprised of several research interests; study of materials' structure; synthesis of nanoparticles, with special attention to magnetic and conductive particles; study and development of coating and adsorption processes.

Manja Kurečič is an Assistant Professor at the Faculty of Mechanical Engineering, University of Maribor. She has 14 years of experiences in the field of nanostructured materials, i.e. nanocomposites (particles and hydrogels) and nanofibrous membranes (electrospun), that can find applications in biomedical area, water purification and other technical applications, as well as their modification and characterization. Her bibliography consists of more than 100 units, among others: 24 reviewed papers, more than 45 conference published lectures and posters, 10 international/national projects reports and 4 invited lectures. In addition, she is a co-inventor of five Slovenian patents.

Andreja Urbanek Krajnc is an Assistant Professor for botany at the Faculty of Agriculture and Life Sciences, University of Maribor. From 1996 to 2004, she studied biology/botany at the Karl Franzens University Graz, Austria. The promotion in Dr. rer. nat. in plant physiology and cell biology at the Karl Franzens University Graz was in 2004. Between 2007 and 2009, she was a postdoc researcher at the Faculty of Agriculture and Life Sciences University of Maribor, Slovenia. Her main areas of research are structural botany, plant cell biology, tree physiology, physiology of agricultural plants (environmental impacts, mineral nutrition, biotic stress and physiological disorders). She collaborates tightly with the experts in the field of horticulture, plant breeding and animal production by analysing metabolites, especially antioxidants and signal molecules as stress markers in plants regarding food safety and health benefits.

Tatjana Kreže is a Professor at the Faculty of Mechanical Engineering, University of Maribor, and the Vice Dean for Quality at the Faculty of Mechanical Engineering. She teaches renewable materials, structure and properties of fibre forming polymers, cellulose fibres, testing of textile materials and standardization. She has published 34 original scientific papers, 2 chapters in a scientific monograph, 1 professional monograph and she has 427 pure citations (Scopus). She is a member of EURATEX—European Technology Platform for the Future of Textiles and Clothing and of Commission for Quality and standardization at Chamber of Commerce and Industry of Slovenia; Technical Committee SIST/TC Textile and textile products and Technical Committee SIST/TC Care labelling of textiles at Slovenian institute for Standardization.

Karin Stana Kleinschek is a Professor at the Faculty of Mechanical Engineering, University of Maribor, where she teaches various courses on polymer chemistry and materials, surface characterization of polymeric materials and textile chemistry. From 2011 to 2015, she was a Vice Rector for Research and Development of the University of Maribor. From 2004 to 2016, she was the Head of the Institute of Engineering Materials and Design, Faculty of Mechanical Engineering, University of Maribor, and is the Head of the Laboratory for Characterization and Processing of Polymers (LCPP), which is a part of the institute. She is a member of various scientific organizations (as Member of the Scientific Committee of International Conferences of Polymer Characterization POLYCHAR, Member of the

Electrokinetic Society Scientific board, Member of the EPNOE-ACS Conference, etc.). She is Vice President for Research of the European Polysaccharide Network of Excellence (EPNOE)—BIC Association. Since 2013, she is a Member of the European Academy for Science and Art and from 2014 Associate Member of Slovenian Academy of Engineering. Since October 2016, she is a Visiting Professor at TU Graz, ICTM. Her fields of expertise are surface modification and characterization of polymeric materials with special attention on polysaccharides and its usability in biomedical applications. Her scientific bibliography consists of more than 950 units (180 peer reviewed scientific papers). She participates and coordinates numerous national as well as international projects. In addition, she acts as thesis adviser and member of organizing committees.

Abbreviations

θ	Scattering angle
λ	Wavelength
ζ-potential	Zeta potential
AMBET	Alkali-assisted microwave plus biological enzymatic technique
BFF	Borassus fruit fibre
CB	Cassava bagasse
CBN	Cassava bagasse nanofibres
CI	Crystallinity index
CMF	Cellulose microfibrils
CNC	Cellulose nanocrystals
CNF	Cellulose nanofibrils
d	Diameter
DLVO	Derjaguin–Landau–Verwey–Overbeek theory
DP	Degree of polymerization
FTIR	Fourier-transform infrared spectroscopy
L	Length
LCA	Life cycle assessment
m_1	Mass of sample after exposure to 65% RH
MAPP	Maleated polypropylene
MCC	Microcrystalline cellulose
MG	Giant Miscanthus
m_o	Mass of absolute dry material
M_s	Moisture sorption
PALF	Pineapple leaf fibre
PDI	Polydispersity index
PHF	Pea hull fibre
PLA	Poly(lactic) acid
PP	Polypropylene
PS	Plasticized starch
PVA	Poly(vinyl alcohol)

RH	Relative humidity
SCB	Sugarcane bagasse
SEM	Scanning electron microscopy
T	Temperature
TEM	Transmission electron microscopy
VC	Variation coefficient
WAXS	Wide-angle X-ray scattering
X_{CR}	Crystallinity index determined by X-ray diffraction
XRD	X-ray diffraction
ZP	Zeta potential

List of Figures

List of Tables

Abstract

Sustainability based on healthy ecosystems and environments is the target idea of today's development, and therefore, it is necessary to follow the same objectives also when studying fibres for new and additional applications; consequently, renewable natural materials with high added value are intensively researched.

An overview of some natural plant sources of cellulose fibres, which have not traditionally been used for the mass production of fibres, is given in the manuscript. Recent research results of fibre properties isolated from non-conventional plants are described. Besides, potential applicability in the field of nanocrystalline cellulose of some of these fibres is reviewed.

In addition, a study of fibres extracted from nettle (*Urtica dioica*), Spanish broom (*Spartium junceum*), lucerne (*Medicago sativa*), weeping willow (*Salix babylonica*), balm-leaved archangel (*Lamium orvala*) and therefrom-derived nanocrystalline cellulose is presented and compared to hemp (*Cannabis sativa*) fibres.

Keywords Sustainable materials · Cellulose fibres · Non-conventional cellulose fibres · Properties · Nettle (*Urtica dioica*) · Spanish broom (*Spartium junceum*) · Balm-leaved archangel (*Lamium orvala*) · Lucerne (*Medicago sativa*) weeping willow (*Salix babylonica*) · Nanocrystalline cellulose

Chapter 1
Introduction

Abstract Different cellulose fibres have been traditionally used for textile and technical applications, however in addition to the traditional cellulosic fibres, fibres isolated from some other non-traditional plants are recently studied for different advanced applications, e.g. cellulose based composites or nanocomposites, etc.

Keywords Renewable fibres · Non-traditional fibres · Composites

In recent years, due to the growing ecological and environmental consciousness, more and more attention is paid to the development of green and renewable materials for different end—use application. Many natural fibres fulfil these conditions perfectly and meet the requirements, both in terms of ecological integrity and natural renewability [1, 2]. Specifically, vegetable cellulose fibres, e.g. stem fibres that occur in primary tissues as bundle sheaths in monocots or bundle caps in dicots, the bast fibers which form fibrous bundles in the inner bark (secondary phloem or bast) of the stems of dicotyledonous plants, the leaf fibres in form of a bundle sheath which run lengthwise through the leaves of monocotyledonous plants and the fibres of seeds and fruits, provide excellent opportunities for different applications [3]. There are about 2000 species of useful fibre yielding plants in various parts of the world and these can be used for many applications [4]. Natural cellulose fibres argued their excellence through history as they accompanied humans for millennia. Green fibres like flax, jute, sisal, kenaf, etc. have been used for more than 8000 years [5, 6].

Despite the development of many new materials with new properties in the last century, the traditional natural cellulose fibres are still among the most important fibres for both apparel and technical textiles. Significant advantages of natural cellulose fibres over synthetic materials (aramid and carbon, etc.) are biodegradability, lightness, low costs, abundance, high biocompatibility, and adequate mechanical properties [7].

Different cellulose fibres can be used for textile and technical applications. Cotton is used mainly for apparel textile and flax, hemp, jute, ramie, sisal and coir for technical purposes. Natural plant fibres will also be the future material, not only

M. Sfiligoj Smole et al., *Surface Properties of Non-conventional*
Cellulose Fibres, Biobased Polymers, https://doi.org/10.1007/978-3-030-10407-8_1

for the textile industry, but also for modern eco-friendly composites used in different areas of application like building materials, automotive and agro materials, particle boards, insulation boards, medicine, and sources for different bio-polymers [6].

Plants, that are traditionally associated with fibres, grow in all geographic areas on Earth, but the climatic conditions limit the growth areas for each species. The most important among cellulose fibres is cotton (*Gossypium* L., Malvaceae); however, cotton cultivation, when examined through the criteria of sustainability, is extremely problematic. Cotton farming involves environmental risks due to intensive use of pesticides and high water demands [8]. The criterion of sustainability is met only by those textiles that include fully renewable materials (raw materials, chemicals, etc.) that, in their production process and their manufacturing, should not consume energy based on fossil fuels. Besides, by definition, sustainable living is taking no more potentially renewable resources from the natural world than can be replenished naturally and not overloading the capacity of the environment to cleanse and renew itself by natural processes [9]. Taking into account the definition of sustainability, just organic cotton meets these requirements for cotton fibres; however, only about 1% of all cotton produced in the world is organic (http://aboutorganiccotton.org/).

In the past, in addition to the traditional cellulosic fibres, fibres isolated from some other non-traditional plants have also had their importance in the field of textile materials. Later, their importance decreased due to economic reasons associated with low yields, worse accessibility and lower supply, or simply because there were enough other natural, and especially synthetic fibres, which met all user needs. At the start of the third millennium, the exponential increase in population increases the demand on food, energy, water, resources and chemicals, and affects the corresponding increase in environmental pollution and a depletion of finite resources (e.g. fossil fuels) [9].

Therefore, manufacture, use and removal of traditional materials are today considered more critically because of increasing environmental consciousness and the demands of legislative authorities, which also require changes in the production, processing, textiles' use and textiles' wastes. One of the very important criteria for materials today is whether they are biodegradable. The material is defined as biodegradable if it is able to be broken down into simpler substances by naturally occurring decomposers. It is an event that takes place through the action of enzymes and/or chemical decomposition associated with living organisms (bacteria, fungi, etc.) and their secretion products [9]. Natural fibres are satisfying this condition perfectly. In general, natural fibres have a great potential as a shift towards the use of more renewable and environmentally friendly materials.

Therefore, we are again exploring almost forgotten sources of fibres, and new plant sources that could complement a series of naturally renewable materials. In this respect, several research projects on various potentially useful plant sources are ongoing. Chemical composition, morphology, structure and properties of these unconventional fibres are studied, as fibres' morphological structure and physico-chemical properties are conditioning fibres' applicability. Above all, research focus is on manufacturing of products of high added value using alternative materials.

Natural cellulosic fibres such, as some specific traditional (e.g. hemp, sisal, jute, flax, etc.) and especially non-traditional fibres, are today applied mostly in composites as reinforcing fibres (e.g. [2, 7, 10–22], etc.). Biocomposites reinforced with natural fibres and/or polymers have developed significantly over the past years. Processing advantages, biodegradability, good mechanical properties, low density, low costs and renewability are decisive therefore [23]. The properties of composites depend on the matrix, fibres and interfacial bonding between the reinforcing fibres and the matrix as the stress transfer between matrix and fibres determines reinforcement efficiency [24]. Because fibres are the reinforcing phase in a polymer matrix, they are influencing the composite mechanical properties positively and reducing the weight of the composite material. Beside several excellent properties, cellulose fibres possess certain drawbacks that impact negatively on the quality of the material for specific applications and reduce their potential as polymer reinforcements. Cellulose fibres are hydrophilic and, consequently, they swell, which is not a desirable feature in a composite material. Secondly, cellulose fibres are not compatible with all matrices; therefore, especially surface morphology, hydrophilic-hydrophobic characteristics and surface charge are of the particular interest in this regard. In order to improve the compatibility of fibres with the matrix a physical and/or chemical pre-treatment procedure of fibres' surface structure is often required prior to composite preparation, as it is known that the incompatibility between the natural material and polymer matrix can influence the strength of the resulting composite (e.g. [25–27], etc.). In order to overcome problems of compatibility and moisture absorption by the fibre during use, several techniques, namely (i) Physical treatments such as corona, plasma, laser, vacuum ultraviolet and γ-radiation treatments; (ii) Chemical grafting by direct condensation, including surface compatibilization with hydrophobic moieties and co-polymerization with the matrix were suggested [28].

Cellulose nano-fibres, nano-crystals and nano-fibrils are advanced cellulose materials with high potential. Cellulose-fibre-reinforced polymer nano-composites have received much attention because of their low density, nonabrasive, combustible, nontoxic, low cost, and biodegradable properties [29]. In addition to traditional sources of cellulose for production of nano-cellulose, non-traditional cellulose sources are also explored and designed as they offer tremendous opportunities in this area.

References

1. Barth M, Carus M (2015) Carbon footprint and sustainability of different natural fibres for biocomposites and insulation material. www.bio-based.eu/ecology. Accessed 20 Dec 2016
2. Mohanty AK, Manjusri M, Drzal LT (2005) Natural fibres, Biopolymers and Biocomposites. CRC Press, Taylor & Francis Group, Boca Raton

3. Sfiligoj-Smole M, Hribernik S, Stana-Kleinschek K, et al (2013) Plant fibres for textile and technical applications. In: Grundas S, Stepniewski A (eds) Advances in agrophysical research InTech, Rijeka, pp 369–397

4. Chirayil CJ, Mathew L, Thomas S (2014) Review of recent research in nano cellulose preparation from different lignocellulosic fibers. Rev. Adv. Mater. Sci. 37:20–28

5. Kozasowski RM, Mackiewicz-Talarczyk M, Allam AM (2012) Bast fibres: flax. In: Kozlowski R (ed) Handbook of natural fibres, 1st edn. Woodhead Publishing, Cambridge, pp 56–113

6. Kozlowski R, Baraniecki P, Barriga-Bedoya J (2005) Bast fibres (flax, hemp, jute, ramie, kenaf, abaca). In: Blackburn R (ed) Biodegradable and sustainable fibres, 1st edn. Woodhead Publishing, Oxford

7. Zhang MQ, Rong MZ, Lu X (2005) Fully biodegradable natural fiber composites from renewable resources: all-plant fiber composites. Compos Sci Technol 65(15–16):2514–2525

8. Cerchiara T, Abruzzo A, Palomino RAŇ et al (2017) Spanish Broom (*Spartium junceum* L.) fibers impregnated with vancomycin-loaded chitosan nanoparticles as new antibacterial wound dressing: preparation, characterization and antibacterial activity. Eur J Pharm Sci 99:105–112

9. Blackburn R (2005) Biodegradable and sustainable fibres, 1st edn. Woodhead Publishing, Cambridge

10. Biagiotti J, Puglia D, Kenny JM (2004) A review on natural fibre-based composites-part I; structure, processing and properties of vegetable fibres. J Nat Fibers 1(2):37–68

11. Fuqua MA, Huo S, Ulven CA (2012) Natural fiber reinforced composites. Polym Rev 52 (3):259–320

12. George J, Sreekala MS, Thomas S (2001) A review on interface modification and characterization of natural fiber reinforced plastic composites. Polym Eng Sci 41(9):1471–1485

13. Hornsby PR, Hinrichsen E, Tarverdi K (1997) Preparation and properties of polypropylene composites reinforced with wheat and flax straw fibres: fibre characterization. J Mater Sci 32 (2):443–449

14. Joshy MK, Lovely M, Rani J (2007) Studies on interfacial adhesion in unidirectional is or a fibre reinforced polyester composites. Compos Interfaces 14(7–9):631–646

15. Kovacevic Z, Bischof S, Vujasinovic E, Fan M (2016) The influence of pre-treatment of *Spartium junceum* L. fibres on the structure and mechanical properties of biocomposites. https://doi.org/10.1016/j.arabjc.2016.08.004

16. Mwaikambo LY (2006) Review of the history, properties and application of plant fibres. Afr J Sci Technol Sci Eng Ser 7:120–133

17. Nekkaa S, Guessoum M, Chebira F et al (2008) Effect of fiber content and chemical treatment on the thermal properties of *Spartium junceum* fiber-reinforced polypropylene composites. Int J Polym Mater Polym Biomater 57(8):771–784

18. Puglia D, Biagiotti J, Kenny JM (2005) A review on natural fibre-based composites—part II; application of natural reinforcements in composite materials for automotive industry. J Nat Fibers 1(3):23–65

19. Rao KMM, Prasad AVR, Babu Babu MNVR et al (2007) Tensile properties of elephant grass fiber reinforced polyester composites. J Mater Sci 42(9):3266–3272

20. Thakur VK, Singha AS, Thakur MK (2012) Green composites from natural fibers: mechanical and chemical aging properties. Int J Polym Anal Charact 17(6):401–407

21. Thakur VK, Thakur MK, Gupta RK (2014) Raw natural fiber-based polymer composites. Int J Polym Anal Charact 19(3):256–271

22. Väisänen T, Haapala A, Lappalainen et al (2016) Utilization of agricultural and forest industry waste and residues in natural fiber-polymer composites: a review. Waste Manag 54:62–73

23. Faruk O, Bledzki AK, Fink HP (2012) Biocomposites reinforced with natural fibres: 2000–2010. Prog Polym Sci 37:1552–1596

24. Sudhakara P, Jagadeesh D, Wang Y et al (2013) Fabrication of Borassus fruit ligno-cellulose fiber/PP composites and comparison with jute, sisal and coir fibers. Carbohyd Polym 98 (1):1002–1010

25. Li X, Tabil GL, Panigrahi S (2007) Chemical treatments of natural fiber for use in natural fiber-reinforced composites: a review. J Polym Environ 15(1):25–33. https://doi.org/10.1007/s10924-006-0042-3

26. Mohanty K, Misra M, Drzal LT (2001) Surface modifications of natural fibers and performance of the resulting biocomposites: an overview. Compos Interfaces 8(5):313–343

27. Zhong J, Li H, Yu J et al (2011) Effects of natural fiber surface modification on mechanical properties of poly(lactic acid) (PLA)/sweet sorghum fiber composites. Polym Plast Technol Eng 50(15):1583–1589

28. Belgacem MN, Gandini A (2005) The surface modification of cellulose fibres for use as reinforcing elements in composite materials. Compos Interfaces 12(1–2):41–75

29. Kalia S, Dufresne A, Cherian BM et al (2011) Cellulose-based bio- and nano-composites: a review. Int J Polym Sci 2011. https://doi.org/10.1155/2011/837875

Chapter 2
Anatomy of Plant Fibres

Abstract Plant fibres are elongated, most commonly sclerenchyma supportive plant cells with thick cellulose walls with a well-organised structure. Cellulose natural fibres are characterised by a fibrillar structure in which the crystalline arranged parts of cellulose I form a periodic array with amorphous cellulose molecules. The morphological hierarchy is as follows: The layers of cell walls are composed of macrofibrils, then of microfibrils and, finally, of finer elementary fibrils. Fibres' morphology and crystalline structure is discussed in this chapter.

Keywords Ultimate plant fibres · Technical plant fibres · Morphology · Crystalline—amorphous structure

The elongated specific shape, high cellulose content in cell walls and thereon-related properties condition the suitability of different plant cells for textile and textile-technical applications. The primary and secondary cell wall determines geometrical shape of the plant cell and its mechanical properties and chemical resistance. Plants cells' walls are formed by a three-dimensional complex network of carbohydrate polymers (i.e. cellulose/ hemicelluloses) embedded in a heterogeneous matrix of pectins and proteins. The other components of the cell wall are secondary metabolites, lignin and mineral. Primary cell wall is found in all cells from algal to terrestrial plants. In some specialised cells such as fibres, vessels, tracheids and sclereids in vascular plants, an additional thicker secondary wall is formed. It is a network of cellulose and lignin [1].

In the research of various plant tissues with the purpose of using them for fibres, biofuels or other technical applications, we must distinguish:

(i) Plant cells are cells of various plant tissues with different geometric forms and different functionality, but often in this context we mean cellulose-rich cells.

(ii) Plant cells with a thick secondary wall contain a high cellulose content, however to emphasize that the main constituents in most plants are cellulose, hemicelluloses, and lignin we use the term lignocellulose, which is

M. Sfiligoj Smole et al., *Surface Properties of Non-conventional*
Cellulose Fibres, Biobased Polymers, https://doi.org/10.1007/978-3-030-10407-8_2

a complex matrix, comprising many different polysaccharides, phenolic polymers and proteins.

(iii) Lignocellulosic biomas is organic material derived from biological origin which has a relatively high content of lignin, hemicellulose, cellulose and pectin combined into a molecular matrix [2].

(iv) Fibers are elongated, most commonly sclerenchyma supportive plant cells with thick cellulose walls. High cellulose content in cell walls is the basis for their useful properties for various applications and also as a source of cellulose.

(v) Textile fibers are fibrous cellulose-rich plant cells, which are isolated by various chemical, biochemical and mechanical procedures from stems, leaves or fruits of different plants. By different textile pre-treatments such as scouring, bleaching, etc., all non-cellulose components are removed from the fibers.

In addition to, numerous studies of cellulosic sources are not performed on isolated cells or fibers, but analyzes are carried out on complete plant structures, such as leaves, stems, etc., which therefore contain all types of plant cells. Therefore, the results obtained are often difficult to compare with each other.

Natural cellulose plant fibres are uni-cellular (e.g. cotton) or multi-cellular elongated structures, located in stems, leaves, seeds or fruits of different plants [3]. Fibres are mainly sclerenchyma cells, which are found almost in all plant parts, including the roots, stem, and in vascular bundles. Wood fibres of trees provide mechanical support and elasticity, bark fibres are important in resisting insects, fungi and other pests. Stem and leaf fibres are mostly phloem fibres, which form large sclerenchyma caps surrounding the phloem and provide mechanical support [4, 5].

Sclerenchyma fibres are spindle shaped elongated, narrow, and thick walled cells with tapering ends providing mechanical support to ground tissues. These cells develop an extensive secondary cell wall that is almost always impregnated with lignin and laid down on the inside of the primary cell wall. Well-differentiated sclerenchyma cells are typically dead at functional maturity as the cytoplasm and nuclei disintegrate, the volume of the protoplast is usually only 5% of the original cell size. But in some species, certain fibres remain alive at maturity and carry out an active metabolism. They are involved in storing starch and Ca oxalate crystals [4, 5].

In Fig. 2.1, cross-sections of plant leaves, stems and seeds are schematically shown. The right side of the picture represent microscopic images of leaves', stems' and seeds' cross-sections. The picture shows that fibrous cells originate from different plant structures, but their common task is to withstand both compressive and tensile stresses in plants.

Figures 2.2 and 2.3 show the cross-section and longitudinal view of a multi-cellular coconut fibre and uni-cellular cotton fibre. Multi-cellular fibres are called technical fibres and represent a bundle of ultimate fibres. Ultimate fibre cells are elongated cells with tapering ends and very thick cell walls, containing a high

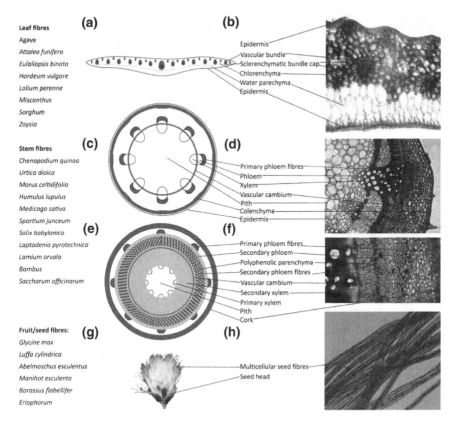

Leaf fibres
Agave
Attalea funifera
Eulaliopsis binata
Hordeum vulgare
Lolium perenne
Miscanthus
Sorghum
Zoysia

Stem fibres
Chenopodium quinoa
Urtica dioica
Morus celtidifolia
Humulus lupulus
Medicago sativa
Spartium junceum
Salix babylonica
Laptadenia pyrotechnica
Lamium orvala
Bambus
Saccharum officinarum

Fruit/seed fibres:
Glycine max
Luffa cylindrica
Abelmoschus esculentus
Manihot esculenta
Borassus flabellifer
Eriophorum

Epidermis
Vascular bundle
Sclerenchymatic bundle cap
Chlorenchyma
Water parechyma
Epidermis

Primary phloem fibres
Phloem
Xylem
Vascular cambium
Pith
Colenchyma
Epidermis

Primary phloem fibres
Secondary phloem
Polyphenolic parenchyma
Secondary phloem fibres
Vascular cambium
Secondary xylem
Primary xylem
Pith
Cork

Multicellular seed fibres
Seed head

Fig. 2.1 The origin of plant fibres. **a** Schematic drawing of a succulent leaf in cross-section representing sclerenchymatic bundle caps as a source of leaf fibres. **b** *Ananas comosus* leaf cross section. **c** Schematic drawing of a primary stem anatomy with bundle caps. **d** Cross-section of *Urtica dioica* stem. **e** Schematic drawing of a secondary stem anatomy representing bark fibres in secondary phloem. **f** Cross-section of a one-year old mulberry (*Morus Alba*) branch representing bark fibres. **g** *Eriophorum vaginatum* seed head with seed fibres. **h** Micrography of *Eriophorum vaginatum* seed fibres

percentage of cellulose. Cells are kidney shaped (cotton), polygonal (e.g. bast fibres) to oval (e.g. ramie) and are enclosed by a cell wall, composed from several layers that are mutually distinguished mainly by their chemical composition, organisation and position of the cellulose microfibrils, while, in the middle of the cell, there is a more or less compressed cell lumen. The cell walls are usually lignified and mineralized providing mechanical support in plants.

In particular, the difference between the cellulose fibres originates from their chemical and physical composition [6]. Common properties of lignocellulose fibres are based on their cellulose origin. These properties are hydrophilicity, biodegradability, good mechanical properties, etc. However, the chemical composition of annual plants, wood fibres, agricultural residues is rather complex and,

Fig. 2.2 SEM and light microscope image of the cross-section and longitudinal view of a coconut fibre

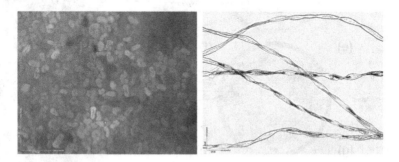

Fig. 2.3 Light microscope image of the cross-section and longitudinal view of cotton fibres

besides cellulose, fibres are composed of several non-cellulosic components such as hemicelluloses, lignin, pectin, proteins, waxes, dyes, etc. Due to less lignin, non-wood plants are a much better option as a cellulose source, since the process of non-wood fibres' delignification and purification is easier and more effective. Besides, fibres from different species differ from each other due to their morphological structure and their properties are strongly dependent on their fibrillar and crystalline structure [7].

2.1   Morphological Characteristics of Technical and Ultimate Plant Fibres

Multi-cellular longitudinal fibre structures are appointed technical fibres and they consist of a different number of elementary or ultimate fibres, held together by pectinous gums, of which the main component is pectins of the middle lamellae and the primary wall. The pectins of the middle lamellae are encrusted with lignin, which makes it even more difficult to split the bundles into elementary fibres [8].

Since mainly the fibres consist of bundles of different numbers of individual cells, and as they are stronger or weaker bonded to other types of plant tissues that surround sclerenchyma fibre cells in leafs or stems (cf. Fig. 2.4), much attention

Fig. 2.4 Cross-section of a hemp stem with fibre bundles

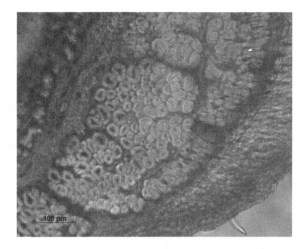

must be paid to the process of isolation and combing. i.e. separation of fibre bundles into smaller ones or single cells. The isolation process influences fibre properties strongly and, consequently, also fibres' usability. Fibres are traditionally retted by dew and water retting [9]. In 14–28 days the pectic materials, that are connecting fibre cells with other cells in stems and leaves, are degraded. In order to avoid long time processes, today's fibres' isolation procedures comprise mechanical decortication and techniques based on chemical, heat and enzymatic degradation of pectins [10].

Different cross-sections' shapes, cell-lengths and finenesses are dependent on the fibre's plant species origin. The width of cellulose ultimate cells ranges between 10 and 50 μm. A cell wall is a composite structure of lignocellulosic material reinforced by helical microfibrillar bands of cellulose (90%) embedded in a matrix of hemicelluloses and lignin [8, 11, 12]. Cell walls are not uniform, but composed of multiple layers of cell walls, which are mutually distinguished mainly by their fibrillar texture and chemical composition. Three cell walls, i.e. primary, secondary and tertiary, are observed in cellulose fibres. During the cell growth on the primary cell wall, which is the outer layer of the fibre, secondary cellulose layers are deposited inwards successively. They are forming the cellulose rich secondary cell wall with three sub-layers (S1, S2 and S3). Each of the three sub-layers has a different microfibrillar orientation or texture [11, 13–15], which is specific for the fibre type. The middle layer consists of helically arranged microfibrils and is the most important for the fibre's mechanical properties.

Cellulose forms mainly the thick secondary cell wall of fibres [7]. The major portion of the non-cellulosic constituents of fibre is present in/or near the primary wall, so the primary wall of the fibre contains almost all of the non-cellulose compounds, except proteins, inorganic salts and coloring matter, which have been found in the fiber lumen [16]. Differences in the contents of various chemical substituents arise from fibre type and, furthermore, from differences in soil, climate,

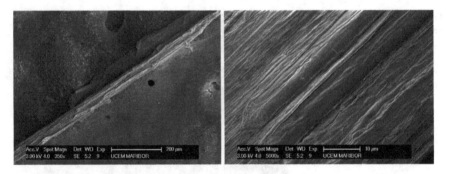

Fig. 2.5 Longitudinal view of a technical fibre isolated from a ryegrass (*Lolium*) stalk (350× and 500×)

weather, farming practices, and other related factors that influence plant growth and fibre development strongly.

In multi-cellular fibres, individual sclerenhyma cells are connected in bundles by sclerenchyma pectin- and lignin-rich middle lamellas (e.g. flax, hemp, ramie, bamboo, and grasses, etc.). Figure 2.5 shows a bundle of connected fibrous cells isolated from a ryegrass (*Lolium*) stalk.

A single technical fibre thus comprises a bundle of ultimates. Sclerenhyma fibre cells are connected to parenchyma cells by parenchyma middle lamellas.

2.2   Crystalline—Amorphous Structure of Different Types of Cellulose Fibres

Cellulose is renewable, biodegradable and non-toxic, and the most abundant natural polymer. The production of cellulose by photosynthesis is estimated to be 1.56×10^{12} t/year [17]. It is found in vascular plants, algae, fungus-like protists (Oomycetes), exceptionally some marine invertebrates (Tunicates) and some bacteria [18].

Cellulose is a high molecular weight polydispersed linear polymer with a syndiotactic configuration. It is based on a 180° turned β-(1,4)-glucopyranoside cellulose polymeric chain. D-glucose building blocks have three hydroxyl groups per anhydroglucose unit. The Degree of Polymerization (DP) is dependent on cellulose origin and is, for natural plant cellulose, up to 20,000 [19, 20]. Purification procedures usually reduce it [6]. It was found that shorter cellulose chains occur in natural plant cellulose as well, however, they are localised mainly in the primary cell walls [20]. The highly functionalized, linear stiff-chain homopolymer is characterised by its hydrophilicity, chirality, bio-degradability, broad chemical modifying capacity, and its formation of versatile semicrystalline fiber morphologies [17, 21].

A number of intra- and intermolecular hydrogen bonds formed by hydroxyl groups present in cellulose macromolecules gives rise to various crystalline domain formations that are considered allomorphs [11, 22]. Various ordered crystalline arrangements are cellulose l, cellulose ll, cellulose lll and cellulose lV. The only native form of cellulose is cellulose I, and it is the native crystalline structure of algae, bacteria and higher plants. The cellobiose segments of two cellulose molecules, one being part of the 002 corner plane and the second being part of the 002 centre plane, house the monoclinic cellulose I unit cell [7, 23]. According to Meyer, Mark and Misch, the dimensions of the monoclinic unit cell are 0.835 nm for the a–axis, 1.03 nm for the b-axis or fibre period, 0.79 nm for the c-axis, and 84° for the ß angle [11]. The model of Meyer, Mark and Misch proposed for a unit cell of crystal lattice assumes a monoclinic cell with the space group P2 and two antiparallel celobiose chain segments running in opposite directions along the fibre axis [24]. The crystalline cellulose I can be converted into the other polymorphs through a variety of treatments [25].

However, the crystalline structure of cellulose is still not understood fully because of its complexity. It is known that the crystalline structure of cellulose I is a mixture of two distinct crystalline forms: cellulose I_α (triclinic) and I_β (monoclinic). Cellulose I_α is reported as the dominant polymorph in primitive organisms (bacteria, algae etc.), while cellulose I_β is predominant in higher plants (woody tissues, cotton, lignocellulose fibres, etc.) [25]. The size of cellulose crystallites is small, generally about 5 nm in width and about 20–30 nm in length [26].

The composition of cell walls is a well-organised architecture. Cellulose natural fibres are characterised by a fibrillar structure in which the crystalline arranged parts of cellulose I form a periodic array with amorphous cellulose molecules. The morphological hierarchy is as follows: The layers of cell walls are composed of macrofibrils, then of microfibrils of 10–30 nm in mean diameter and, finally, of finer elementary fibrils of 3 nm in diameter [7, 19, 27].

Crystalline domains, which typically represent the majority of the fibrils, demonstrate a typical X-Ray Diffraction XRD pattern characteristic of cellulose I. A typical natural crystalline cellulose XRD diagram represents equatorial diffraction peaks at the 2θ angles of about 14° and 16° and the strongest diffraction peak at an angle of 22° [28]. There is an additional peak at 34°. These peaks are the characteristic positions of distinct and independent reflections 101, 10-1, 002, and 040 [23]. The non-crystalline part of the cellulose structure is represented by broader and less clearly refined features in the diffraction pattern [26].

However, after an alkaline treatment with NaOH with a concentration over 12%, two additional diffraction peaks appear at 12.2° and 20.1°, that are assigned to cellulose II. The greatest efficiency of polymorphic transition of cellulose I to cellulose II is obtained at 16% NaOH concentration [28].

The fraction of the ordered crystalline structures in the polymer is expressed by the crystallinity index (CI). Although measurements of crystallinity index have a long history, it has been found that CI varies significantly depending on the choice of measurement method [29]. Mostly measuring methods assume an ideal polymer mixture of perfect crystalline and totally disordered areas, which ignores the transition structures. In addition to, different measuring techniques are based on totaly different physical and physico-chemical theoretical bases. Usually, crystallinity of polymers is determined by density measurements, measuring the melting enthalpy, by infrared spectroscopy, nuclear magnetic resonance, and most precisely by X-ray scattering, although also by x-ray analysis the technique itself (i.e. the measurement geometry, the evaluation of diffractograms, etc.) strongly influences the result.

In addition, crystalline phase analyzes often take place on differently treated materials, which means that the results given in literature can refer to pure cellulose, isolated fibers with present non-cellulose components, or even to plants' parts such as leaf and stalk fragments. Consequently, high differences in the crystallinities given in the literature, are detected. Common cellulosic materials have crystallinities in the range of about 26–80% [30].

References

1. Costa G, Plazanet I (2016) Plant cell wall, a challenge for its characterisation. Adv Biol Chem 6:70–105. https://doi.org/10.4236/abc.2016.63008
2. Wu S, McLaren J, Madl R, Wang D (2009) Biofuels from lignocellulosic biomass. In: Sustainable biotechnology: sources of renewable energy, pp 19–41. https://doi.org/10.1007/978-90-481-3295-9_2
3. Cook JG (1993) Handbook of textile fibres: natural fibres. Merrow Publishing, Durham
4. Bresinsky A, Körner C, Kadereit W et al (2008) Strasburger—Lehrbuch der Botanik. Spektrum, Heidelberg
5. Nultsch W (2001) Algemeine Botanik. Thieme, Stuttgart
6. Bledzki AK, Gassan J (1999) Composites reinforced with cellulose based fibres. Prog Polym Sci 24:221–274
7. Lewin M, Pearce EM (1998) Handbook of fibre chemistry, 2nd edn. Marcel Dekker New York, Basel
8. Kostic M, Pejic B, Skundric P (2008) Quality of chemically modified hemp fibers. Bioresour Technol 99:94–99
9. Östbom G, Svennerstedt B (2007) Fiber quality of hemp grown on the Swedish island Gotland. In: Hunter L (ed) Anandjiwala R. Textiles for sustainable development, Nova Science Publishers Inc., pp 53–61
10. Paridah T, Basher AB, SaifulAzry S et al (2011) Retting process of some bast plant fibres and its effect on fibre quality: a review. BioResources 6(4):5260–5281
11. Krässig HH (1992) Cellulose, structure, accessibility and reactivity. Gordon and Breach Science Publishers
12. Li Y, Mai YW, Ye L (2000) Sisal fibre and its composites: a review of recent developments. Compos Sci Technol 60(11):2037–2055
13. Cuissinat C, Navard P (2008) Swelling and dissolution of cellulose, part III: plant fibres in aqueous systems. Cellulose 15(1):67–74
14. Eder M, Burgert J (2010) Natural fibres—function in nature. In: Müssig J (ed) Industrial applications of natural fibres: structure, properties and technical applications. Wiley, Chichester, pp 23–41
15. John MJ, Anandjiwala RD (2008) Recent developments in chemical modification and characterization of natural fiber-reinforced composites. Polym Compos 29(2):187–207
16. Bismarck A, Aranberri-Askargorta I, Springer J et al (2002) Surface characterization of flax, hemp and cellulose fibers; surface properties and the water uptake behavior. Polym Compos 23(5):872–895
17. Klemm D, Heublein B, Fink HP et al (2005) Cellulose: fascinating biopolymer and sustainable raw material. Angew Chem Int Ed 44(22):3358–3393

18. Nechyporchuk O, Belgacem MN, Bras J (2016) Production of cellulose nanofibrils: a review of recent advances. Ind Crops Prod 93:2–25
19. Brinchi L, Cotana F, Fortunati E et al (2013) Production of nanocrystalline cellulose from lignocellulosic biomass: technology and Applications. Carbohyd Polym 94:154–169
20. Habibi Y, Lucia LA, Rojas OJ (2010) Cellulose nanocrystals: chemistry, self-assembly and applications. Chem Rev 110(6):3479–3500
21. Klemm D, Kramer F, Moritz S et al (2011) Nanocelluloses: a new family of nature-based materials. Angew Chem Int Ed 50:5438–5466
22. Hubbe MA, Rojas OJ, Lucia LA et al (2008) Cellulosic nanocomposites: a review. Bioresources 3:929–980
23. Hu XP, Hsieh YL (1996) Crystalline Structure of Developing Cotton Fibers. J Polym Sci Part B Polym Phys 34:1451–1459
24. Wüstenberg T (2015) Cellulose and cellulose derivatives in the food industry: fundamentals and applications. Wiley-VCH, Weinheim. https://doi.org/10.1002/9783527682935.ch03
25. Sun JX, Sun XF, Zhao H et al (2004) Isolation and characterization of cellulose from sugarcane bagasse. Polym Degrad Stab 84(2):331–339
26. Terinte N, Ibbett R, Schuster KC (2011) Overview on native cellulose and microcrystalline cellulose I structure studied by X-ray Diffraction (WAXD): comparison between measurements techniques. Lenzinger Berichte 89:118–131
27. John MJ, Thomas S (2008) Biofibres and biocomposites. Carbohydr Polym 71:343–364
28. Yueping W, Ge W, Haitao C et al (2010) Structures of bamboo fiber for textiles. Text Res J 80(4):334–343. https://doi.org/10.1177/0040517509337633
29. Park S, Baker JO, Himmel ME et al (2010) Cellulose crystallinity index: measurement techniques and their impact on interpreting cellulase performance. Biotechnol Biofuels 3:10. https://doi.org/10.1186/1754-6834-3-10
30. Hubbe MA, Gardner DJ, Shen W (2015) Contact angles and wettability of cellulosic surfaces: a review of proposed mechanisms and test strategies. BioResources 10(4):8657–8749

Chapter 3
Non-conventional Plant Fibres

Abstract There are many non-conventional fibre-yielding plants with the potential for use for different applications. Some of them are presented in the chapter.

Keywords Fibres from wild growing plants · Fibres from waste crops · Quinoa · Nettle · Mulberry · Hop · Soy · Lucerne · Spanish broom · Loofah · Okra · Babylon willow · Cassava · Khimp · Balm-leaved archangel · Agave · Piassava · Palmyra palm · Cotton grass · Bamboo fibres · Sabai grass · Barley · Ryegrass · Giant miscanthus · Sugarcane bagasse fibres · Sorghum · *Zoysia*

There are many non-conventional fibre yielding plants with the potential for use for different applications, however mainly they are unexplored so far [1]. The focus of studying non-traditional plant fibres is on developing new environmentally friendly materials and processes for the use of biomass, which can replace classical procedures that are based on the exploitation of petroleum resources gradually. These materials are a promising reinforcement for use in composites on account of their low cost, low density, high specific strength and modulus, no health risk, easy availability in specific areas and renewability.

Lignocellulosics are used for various applications, depending on their composition and physical properties. Usually, these materials are already utilised as cellulose sources, e.g. wheat and rice straw, corn stalks in pulp and paper production, as well as for textile uses (e.g. fibres from coir, pineapple and banana leaves). Recently, other lignocellulose fibers from additional non-tradition origins have also been examined and new opportunities for their use are being studied [2].

Non-conventional cellulose fibres can be divided into two groups. The first originates from plants which grow wild on non-agricultural land. The advantage of these materials is also their short renewal time. Plants for non-traditional fibres' extraction do not occupy cultivated land and, generally, can be cultivated without the use of pesticides. Provided that these plants are widespread, their exploitation represents a good opportunity for obtaining cellulose fibres.

These plants grow mainly in tropical and subtropical areas in addition to grasses, which are more or less prevalent in all geographical zones. The exploitation of

M. Sfiligoj Smole et al., *Surface Properties of Non-conventional*
Cellulose Fibres, Biobased Polymers, https://doi.org/10.1007/978-3-030-10407-8_3

certain non-traditional plants for isolation of cellulose fibres would, for some countries, especially non-developed countries, represent significant economic benefits.

The second group consists of waste crops that do not currently have specific applicability and are removed by burning or waste composting. Their removal causes impacts on the environment and, therefore, their use in high added value products will bring even more benefits. Agro-based bio-fibers have the composition, properties and structure that make them suitable for uses such as composite, textile, pulp and paper manufacture [2]. A broad spectrum of fibres can be made from by-products salvaged from industrial, agricultural, or commercial production. There are several different agricultural residues with very high potential for new renewable products, and some of them are already in the phase of the transfer into practice, whilst, for others, their properties and potential utilization is being researched. Some biofibres from agricultural by-products are wheat straw, pineapple leaves, bamboo, sugarcane bagasse, hop stems, banana shoots and leaves, quinoa, okra, etc. In addition to, artichoke (*Cynara cardunculus*), *Ferula*, *Althaea*, *Sansevieria* and *Arundo donax* were studied and characterized as new natural fibres [3]. Morphological characteristics (length, width, etc.) of fibers obtained from different plant species are highly dependent on growing conditions and plant biology [4]. Furthermore, physico-chemical and mechanical properties, surface chemistry and thermal degradation characteristics of fibres are significantly influenced by the fibres' isolation process [5].

High demand of natural resources due to rapid urbanization and the disposal problem of agricultural wastes in developed countries have created opportunities for use of agro-waste in the construction industry. Many agricultural waste materials are already used in concrete as replacement alternatives for cement, fine aggregate, coarse aggregate and reinforcing materials. Different agricultural wastes were involved, i.e. groundnut shell, oyster shell, cork, rice husk ash and tobacco waste, bagasse ash, sawdust ash, etc. [6].

In addition to these, non-traditional fibres are being studied for their utilization as sustainable and eco-friendly acoustic absorbers to replace synthetic materials which are still applied extensively in the building industry. For sound absorbance, good acoustic performance was found for samples from sugarcane wasted fibres. At 1.2–4.5 kHz, the average absorption coefficient is 0.65, which is comparable against that from the classical synthetic absorbers [7].

Some non-traditional plant species for fibres' extraction that have been studied recently with the aim of extracting fibers are presented in this chapter. The plants are systematically arranged to dicotyledons and monocotyledons in the order of lower taxonomic units (order, family).

3.1 Dicotyledonous Fibre Plants

3.1.1 Quinoa

Quinoa (*Chenopodium quinoa* Willd.) is a pseudocereal belonging to the amaranth plant family (*Amaranthaceae*) with a very long tradition of cultivation. It originated in the Andean regions of Peru, Bolivia, Ecuador, Columbia and Chile, and was domesticated 5000 B.C. as a cherished crop of the Incas. It was the most important food in that region when Europeans arrived. The Spanish Conquistadors, in an effort to subdue the Incas prohibited the cultivation of quinoa. The crop is now achieving a minor comeback internationally defined as one of the "grains of the 21st century" and studied intensively as a health food [8–10]. It is a dicotyledonous annual plant, usually about 1–2 m high. The fruit is an achene, which matures in 5–6 months. The stem is woody and branched. It may be green, red or purple. The plant has deep penetrating roots and is notable for its adaptability to all types of soils and climates. It can be cultivated from sea level up-to an altitude of 3800 m, replacing maize. Its cultivation has spread to every continent. The plant shows tolerance to frost, salinity and drought, and has the ability to grow on marginal soils [11, 12].

Quinoa fibres were studied by Sfiligoj et al. [13]. A mean diameter of 164 μm was determined from SEM images for fibre bundles isolated from untreated stems; however, pre-treatment of stems reduces the fibres' diameter to 42.61 μm. Different methods with different efficiency in the separation of fibrous and non-fibrous stem tissues were used for fibres' extraction. Fineness of fibre bundles was between 24.66 and 96.84 dtex depending on the isolation method used for fibres' extraction [13]. Figure 3.1 shows the surface morphology of isolated fibres.

The stress–strain behaviour of plant structures indicates a rigid characteristic with low elongation and high breaking strength. The elongations vary between 1.4 and 2.1% and tenacities from 13.78 to 32.19 cN.tex^{-1}, which is comparable to some other textile bast fibres, although the high brittleness of dry quinoa fibres is problematic for its utilization [14].

Fig. 3.1 SEM image of quinoa fibres isolated by alkaline treatment (1% NaOH at a temperature of 100 °C; treatment time 1 h)

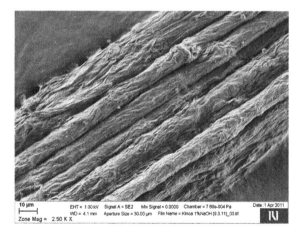

3.1.2   Nettle

The nettle, *Urtica*, is a perennial flowering plant genus in the family Urticaceae, which brings together 30–45 species [15]. Stinging nettles' (*Urtica dioica* L.) habitat is widespread in throughout the temperate regions of Europe, Asia, America. It is a perennial, nitrophilous herb that grows very easily and reaches a height of up to 1.20 m [15]. Many hollow stinging hairs on leaves and stems called trichomes are typical [16]. It grows on phosphates and nitrates rich soils, preferably on soils with adequate moisture and pH 6.5 in shady habitat. In favourable conditions it grows rapidly. Nettle is considered a weed and is resistant to pests and, therefore, does not need pesticides even if it had been cultivated deliberately.

Because of the high plant abundance and good fibre properties, nettle fibres were used for rope-making and fabric production a long time ago. Nettle cloth was comparable to linen cloth or even more fine. There is evidence that the production of textiles from nettle fibres continued to the 17th century in Scandinavia, Poland, Germany and Russia, in Poland even to the 19th century. Because of the great lack of cotton in the first half of the 20th century, fibres from nettles were used to produce uniforms, tents, backpacks, socks, etc. Green unbleached fibers also served as camouflage [17]. Later after the Second World War, nettle cultivation was abandoned due to technical and economic reasons, and especially due to the significant impact of the cotton industry [18].

Stinging nettle has particularly a long history as a fiber plant in Germany and Austria [19]. In the forties and fifties of 20. century superior genotypes were selected by Bredemann [20] at the University of Hamburg, which were characterized by high fiber content, up to 16% compared to around 4–5% in wild nettle, long and not ramified stalks and strong tillering. The most interesting genotypes are still stored at research institutes in Germany.

Figure 3.2 shows a cross-section of the nettle stem with fibre bundles between parenchyma tissues.

Recently nettle morphology and properties were studied by several researchers. It was found that plant stems are reinforced by fibrous cells, which are comparable to flax and hemp fibres. Different nettle retting procedures were studied, i.e. chemical extraction, water retting, microbiological and enzymatic methods to evaluate morphological and mechanical properties and chemical composition of isolated fibres [21]. Bast content in nettle straws is about 20% and it is less than in flax and hemp [22].

Fibres' chemical composition is 54% cellulose, 10% hemicelluloses, 4.1% pectin, 9.4% lignin, wax and fat 4.2%. The Degree of Polymerization (DP) of cellulose in cell walls is 2280 [23]. Fibres are resistant to different media. Elementary fibres are white, silky, and can reach a length of up to 5 cm. Technical fibres, which are bundles of elementary ultimate fibres, are up to 90 cm long and 40–125 µm wide; however, for ultimate fibres, it was determined that fibre mean diameter values ranged from 47 to 19 µm and fibre length from 43 to 58 mm moving from stalk bottom to top [24].

Fig. 3.2 Cross-section of a nettle stem

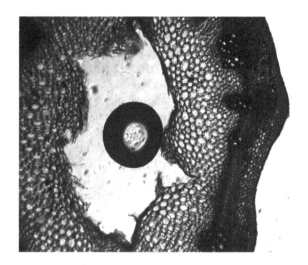

The cross-section is a polygonal with rounded edges. The fibres are twisted. An intrinsic heterogeneity was observed of the fibres along the stem. These heterogeneities influence the chemical, physical and mechanical properties of fibres extracted from different portions of stalks. Fibres' tenacity varies from 24 to 60 cN tex^{-1} for fibres from the bottom part and fibres from the upper part of the stalk, respectively. The elongation of fibres is 2.3–2.6% [24]. Specific weight of nettle fibres is 0.72 g cm^{-3} [24].

The mechanical properties of extracted fibres are better than cotton; however, fibres are finer than hemp; they are soft, fire retardant and are biodegradable. Recently, stinging neetle has been reintroduced in Germany as a collection of nettle fabrics with the trade name Nettle World, confirming the evidence that the use of nettle for a sustainable and local fiber-producing industry, with low environmental impact, could be economically promising [25–27]. In addition, in term of growth and fibre yield in different environments, various genotypes have been selected over the last two decades [24].

3.1.3 Mulberry

Mulberries are deciduous trees in the genus *Morus*, which belongs to the plant family Moraceae. The white mulberry (*Morus alba* L.) origins from south and south-western China where it has been widely cultivated for more than 4500 years mainly for leaf yield in sericulture to feed the caterpillars of the silk producing Lepidopteran insect *Bombyx mori* L. The white mulberry was brought to Europe in the 11th century together with sericulture. Today white mulberry trees can be found all over the Europe, especially in southern former silk-producing countries. In European countries it is mainly grown for fruit production and it is also used as vegetable and feed for animals in different parts of the World. Wide range of medicinal activities have been credited to the different parts of the mulberry plant [28].

The bark of mulberry tree is fibrous and is traditionally used for paper-making and textile fiber extraction. Bark paper (known as amate in Mexico) is part of the history of Mesoamerican cultures, closely embedded in religious, sacred, and political life. It is supposed that amate paper making is of Mayan origin, approximately 300 B.C. [29]. In the past, amate [Ficus tecolutensis (Liebm.)Miq] and moral (Morus celtidifolia Kunth). were the species most commonly used for bark paper, however the majority of the bark paper currently produced is made from the bark of Trema micrantha (L.) Blume [29, 30].

In contrast to paper making, mulberry bark was the native American peoples' fiber source. They extracted the remarkable fibers from the inner bark or secondary phloem and used them for their clothing and everyday needs. Later, other materials like sheep's wool and cotton, both nonnative fibers, were introduced through exchange with European settlers or other native tribes. The writings of many early European travelers describe a fantastic cloth made of mulberry bark. Regardless, the mulberry fiber was not the only source of bark fiber used. Bark fibers from the slippery elm (Ulmus rubra Muhl.) or basswood (Tilia) were used by Southeastern tribes while the Choctaw word 'baluhchi' refers to the hickory (Carya) bark used in making ropes [31].

The first attempts to use mulberry bark for textile fibers' production in Italy trace back to the 19th century. However, mass utilization of mulberry bark is not carried out yet. The textile fiber extraction gained increased scientific attention in recent years [32, 33].

Li et al. firstly obtained cellulose fibers from mulberry branch-barks by high temperature alkali treatment [33]. Fibres were then hydrolyzed by sulfuric acid to produce cellulose nano-whiskers. Generally, the cellulose fibers obtained at 130 °C by alkali treatment had higher cellulose content and lower hemicellulose and lignin contents than those obtain at 80 °C, which had the size predominantly ranged from 20 to 36 nm and high thermostability. On the other hand, the cellulose whiskers produced by sulfuric acid hydrolysis had high crystallinity. Crystallites' size ranged from 20 to 40 nm and their length from 400 to 500 nm. The cellulose whiskers revealed two-stage decomposition behavior due to the introduction of sulfated groups into the crystals in the sulfuric acid hydrolysis process [33].

Qu and Wang [34] successfully obtained mulberry fibers by a new pretreatment named alkali-assisted microwave plus biological enzymatic technique (AMBET), which successfully removed the impurities (such as hemicellulose, lignin, pectin and aqueous extract) of the bast fibers. AMBET treated mulberry fibers were even, smooth and fine, and typical cellulose I allomorph in the mulberry fibers was confirmed by FTIR and XRD analysis. The crystallinity of the AMBET treated fibers was higher than that of the raw mulberry and chemical treated mulberry fibers. Thermal analysis indicated that the mulberry fibers had a good thermal stability. Moreover, the AMBET treated mulberry fibers showed excellent antimicrobial activities against Staphylococcus aureus. The bacteriostatic rate of both the chemical treated and AMBET treated mulberry fibers was higher than 60%, which indicated that the treated mulberry fibers had excellent antimicrobial activities due to some antimicrobial compounds such as tannic acid, flavone and phenols, which have antimicrobic activity against to the *S. aureus*. Moreover, the treated mulberry fibers had a higher bacteriostatic rate, which was due to the removal of the pectin during the treatment processes since the pectin could supply nutrient for *S. aureus*. Mulberry fibres' physical properties indicate that AMBET treated fibers would be ideal candidates for new textile materials [34].

Recently, Dong et al. extracted fibers from the bark of mulberry branches with low gum content by a combination of bacteria and peroxide treatments. The bark of mulberry branches, with 30% cellulose, is a copious and inexpensive source of natural fibers. However, fibers extracted by microwave, enzyme or alkali had a high gum content, which rendered them difficult to be made into high-value textiles. Bacteria strains with high polygalacturonase activities and subsequent hydrogen peroxide treatment decreased the hemicellulose content to 2.5% and lignin content to 2.4%. Compared to flax yarns, mulberry fiber yarns had tenacity 20% higher, elongation 18% higher and unevenness 30% lower than flax yarn. Cotton/mulberry fiber fabrics had softer and smoother hand than cotton/flax fabrics. Overall, the fibers in the study of Dong et al. showed better potential for industrial textile applications than those in previous studies [32].

Efforts have also been made in recent times for sustainable dyeing and value-added finishing of textiles using mulberry fruit extract mainly from black mulberry (*Morus nigra* L.) and red mulberry (*Morus rubra* L.) fruits which has a strong staining activity containing natural anthocyanin dyes (mainly consisting of cyanidine 3-glycoside) with a dyeing property performance good enough for potential commercial applications [35, 36].

3.1.4 Hop

Hop (*Humulus lupulus* L.) is a perennial dioecious climber of the Cannabaceae family, which produce clusters of either female or male flowers. Female inflorescences possess numerous glands with volatile oils that produce the characteristic flavors. They are mainly used to impart bitterness, flavour and preservation properties in beer production and herbal medicine. Wild hops are widely distributed throughout the Northern Hemisphere; i.e. Europe, Asia and North America [37, 38]. Commercial hop production occurs in many countries. Hop cultivars have been roughly classified into two groups, one group consisting of European landraces while the other has a genetic composition derived from North American wild hops [39, 40]. The history of hops and their usage is surrounded by uncertainty. How hops were domesticated in Europe is also unknown although pollen records suggest hop usage had already started in the early Roman era [41–43]. In the 10th century stems were used as bedding and insulation, and twine was made from their fibers. Both the stems and inflorescences were used at that time as packing material in England. Furthermore, the stem fibres were used to produce a coarse cloth [44].

A hop vine can grow to a height of over 7 m. The very long stems wither at the end of the growth season and are agricultural waste [45]. Hops stems were analysed for fibres extraction. For the fibres, obtained by alkaline extraction from the bark of hop stems high cellulose content, low crystallinity and high crystalline orientation was determined. Mechanical properties of fibres were comparable to properties of hemp fibres, i.e. the strength and modulus of the fibers were lower but elongation was higher when compared to hemp [46].

3.1.5 The Fibre Plants of the Pea Family

3.1.5.1 Soy Hulls

Soy hulls are an agro-industrial residue available in huge quantities throughout the world. Soy [*Glycine max* (L.) Merr.] is a species of a large pea family (Fabaceae) native to East Asia, grown widely for its edible beans, which have numerous uses. It is believed to have originated in northern China and was first cultivated there in at least 3000 B.C. The plant is classed as an oilseed and became since 1970 the world's primary protein plant.

Harvesting soybeans generates a lot of agricultural residues, i.e. leaves, stems and pods. Soy hulls are obtained in the soybean oil extraction process, since it is necessary to remove the film coating the grain for the oil extraction [47]. Soybean harvesting residues are usually disposed as waste through landfilling, incineration, or dumping. However, soybean residues are cellulose-rich materials, which contain a sufficient amount of cellulose to justify their use for more advanced applications. Cell morphology, chemical components, lignin distribution and inorganic elements distribution of corn stover fractions, etc. was studied by several authors [e.g. [48–51].

Chemical composition of soybean straw is: 35% cellulose, 21% insoluble lignin, 17% hemicelluloses, 11% ash, and 1% acid soluble lignin, protein, pectin, and glucuronic acid substitutes. The composition of stems and pods is different: The content of holocellulose is 69.2 and 53.8%, respectively, the lignin content is 21.6 and 17.2% respectively, ash content is 2.28 and 7.25%, respectively [48, 51]. In addition, it was confirmed that, similar to other cellulosic materials, the content of cellulose increases with different treatments. Alkali treatment and bleaching raise the cellulose content from 39.8% for untreated soybeam straw to 66.2% for treated straw (NaOH 17.5%, 15 h/30 °C and H_2O_2 4%, 3 h/90 °C), and the amount of hemicellulose and lignin decreases consequently [51]. The average length, width, and length-width ratio of soybean stem fibres are 0.46 mm, 24.2 μm and 19, respectively [50].

After initial treatment of soy hulls by alkali, which is performed to solubilize lignin, pectins, hemicelluloses and proteins, and bleaching to remove the lignin residues, the fibre content in hulls is about 4–6% (w/w). The contents of cellulose are 48.2 ± 2.1% and 84.6 ± 4.0%, of hemicelluloses 24.0 ± 3.0% and 11.2 ± 4.0%, and of lignin 5.78 ± 1.06% and 3.67 ± 0.33%, for untreated and treated soy hulls, respectively. The index of crystallinity increased from 26.3% for a raw sample to 67.2% for an alkali treated and bleached sample [47].

Nanocrystals were prepared successfully from purified soy hulls by acid hydrolysis. Aggregates of nanocrystals and individual needle-like particles were obtained and analysed [47].

3.1.5.2 Lucerne

Medicago sativa L. (lucerne, alfalfa) is the world's most widespread and important forage legume. Although the cultivated alfalfa is tetraploid, primitive diploids occur in Iran and the other parts of the Near East, believed to be the area of original domestication. Alfalfa is a perennial with up to 1 m height, the normal growth period from four to eight years depends on variety and climate. It can be used as a pasture crop or harvested and used as hay and silage. Harvesting of lucerne is three or more times per year, in warm regions with adequate precipitation, as many as six to nine cuttings. Growing lucerne requires fertile and well-drained soil with a pH close to neutral. Lucerne develops a comprehensive, well-branched root system that can penetrate deep into the soil. In good quality soil roots can grow to more than 6 m deep. Enhanced root system enables lucerne access to water and nutrients. Lucerne forms a symbiotic relationship with nitrogen-fixing bacterium *Sinorhizobium meliloti*. The cultivation of lucerne improves the quality of the soil, increases the organic substances and promotes the penetration of water into the soil [52, 53].

A xylem tissue with thick cell walls is developed in the stalks. It consists of cellulose, hemicellulose, pectin and lignin. A mature plant contains about 30% cellulose, 12% hemicellulose, 12% pectin, and 17% lignin, 8.8% protein, 6% ash, 2% of starch, and 12.2% other substances (lipids, carbohydrates) [54]. The mature plant develops a thick secondary cell wall particularly rich in cellulose. The average fibre length is 0.78 mm [55], although there are some differences between a fibre's length in stems and leaves (1.2 and 0.6–1.3 mm, respectively). Fibers' bundles fineness is 19–65 dtex [56].

3.1.5.3 Spanish Broom

Spanish broom (*Spartium junceum* L.) is a perennial 1–1.5 m tall shrub from the pea family (Fabaceae) growing in hot and dry climate throughout the Mediterranean area, in South West Asia and North Africa, where it naturally occurs in hilly limestone and salty soils. In Italy it grows up to an altitude of 975 m, in Turkey, Syria and Palestine even higher up to 1700 m. In the west coast of North America and New Zealand, it is defined as a weed. It contributes to lower erosion and risks of nutrient leaching [57–62].

As a legume plant, *Spartium* uses symbiosis with bacteria in roots to bind atmospheric nitrogen and, thereby, increases the concentration of nitrogen compounds in the soil. Spanish broom blooms with intense yellow flowers from May to

July, the seeds ripen from August to October. The usefulness of the plant is doubled, for the extraction of fibres and seed oils. The percentage yield of the oil ranges from 4 to 5% [63]. Broom is xerophile and a number of adjustments to water shortages are visible in the microscopic structure of vegetative organs. The upper part of the stem, phylocladium, assumes the function of a plant leaf, while sclerenchyma fibers occur cortical and around bundles [57–62]. The name *Spartium* is from the Greek word denoting "cardage," in allusion to the use of the plant. Fibres were known already by the Greeks, Romans and Carthaginians, and used for manufacturing of ropes, nets, bags, sails, clothing. The use of Spanish broom fibres for textile materials was abandoned due to the difficult fibres' extraction process. During World War II, because of the lack of flax, hemp and cotton in Italy, they again started to use Spanish broom. Today, there are a few manufacturers of traditional crafts using Spanish broom fibres [64].

The main difficulty in obtaining fibres is represented by the removal of the epidermis and cuticle, consisting of alkenes, alcohols, esters, and removal of the woody parts, and pectin [65].

Spanish broom cortical fibers are multiple elementary fibers of high lignin content surrounding the bundles. 8–12% of fibres are extracted from plants. Ultimate fibres are regular in shape. The diameter of the elementary fibres varies from 5 to 10 μm, while the diameter of the whole bundle is about 50 μm [64, 66]. The length of technical fibres in stems is up to 40 cm [67]. A thick secondary cell wall indicates a high content of cellulose. Spanish broom fibre chemical composition is approximately 45% cellulose, 25% lignin and 25% of other substances and 5% ash [66]. Due to the high lignin content the untreated fiber is less hygroscopic. The cellulose Degree of Polymerization of Spanish broom fibres is about 1700 and the index of crystallinity 52% [66].

Fibres are characterised by excellent mechanical properties. They are very rigid with the load-elongation diagram almost linear up to the fracture. Fibres' tensile strength is 700 MPa, elastic modulus 20 GPa, which is comparable with the modulus of E-glass fibres (70–90 GPa) [66].

Different methods have been studied for the extraction of cellulose fibers from Spanish broom. In recent years, pectinolytic enzymes that degrade the pectin of the middle lamella leading to separation of the cellulose fibers produced by micro-organisms have gained most attention [64]. A novel method for obtaining fibres of high quality is suggested by Gabriele et al. The method is based on a combination between an initial alkaline treatment, followed by compression with hot air in an autoclave and rapid decompression. Fibres obtained by this isolation process have excellent physical–chemical properties, such as high mechanical resistance and high elasticity [58]. Fibres are well separated and with diameters between 7 and 10 μm. The length of elementary fibers is between 25 and 27 mm [58].

Different uses have been proposed for Spanish broom fibres. Due to the exceptional resistance of fibres to weather conditions it was proposed to use Spanish broom fibres for preparation of painting canvases. As artworks, and especially paintings, appear very sensitive to environmental conditions, the degradation of paintings is one of the problems in the field of cultural heritage and, therefore,

Spanish broom canvas, because of its excellent properties, is proposed as a novel painting support [68].

With the aim to prepare sustainable construction materials adobe bricks were realized with natural material locally available in the Mediterranean area. Bricks were tested with different kinds of reinforcement fibers, i.e. straw and Spanish broom, which spontaneously grows all over the Mediterranean basin. The impact of Spanish broom fibers on the compression strength of the adobe bricks was positive because of the high tensile strength of the fibers and their adhesion to the clay matrix [69].

Attempts were made to prepare composites with either degradability or fire-proofing properties with Spanish broom fibres. Spanish broom fibres' derivatives were prepared by introducing polymerizable or polyhalogenated residues in pre-treated fibers. The formation of derivatives was confirmed by different methods. The application of modified fibres is in agriculture and for packaging materials [70].

A new wound dressing based on Spanish broom fibres impregnated with van-comycin loaded chitosan nanoparticles was proposed by Cerchiara et al. [71]. Fibres obtained from Spanish broom were used for wound dressing preparation because of the low cost of fibres, their availability and sorption properties. Fibres are extremely hydrophilic. In addition, their mechanical properties and their mor-phology meet the requirements for materials for wound dressings perfectly [71].

3.1.6 Loofah

Loofah [*Luffa cylindrica* (L.) M. J. Roem.] is a tropical annual plant belonging to the family of Cucurbitaceae with ancient cultivation in the Old World. It is now cultivated in all tropical regions of the world as vegetable and for yielding loofah sponges. Although of tropical origin, the best loofahs are grown in Japan. Fruits possess a netting–like fibrous vascular system. Before World War II loofahs were mainly imported to United States in order to use them for filters in marine steam engines and in diesel engines. They are also used as bath sponges and materials for adsorption in water treatments, cleaning purposes and in the manufacture of potholders, mats, insoles, sandals and gloves. Because of their shock and sound absorbing properties they were used in steel helmets and armoured vehicles of the U.S. Army. Recently, it was studied as a filler in composites [53, 72, 73, 74]. Loofah fibers also promise being a good adsorbent for removing dyes from aqueous solution since they are renewable and sustainable and show high adsorption capacity and BET surface area [75].

The chemical composition of isolated fibres depends of plant genotype and environmental conditions. Siqueira et al. determined the content of cellulose ($65.5 \pm 0.5\%$), hemi-cellulose ($17.5 \pm 0.5\%$), lignin ($15.2 \pm 1.0\%$) and extrac-tives ($3.1 \pm 0.5\%$). A high index of crystallinity was determined for cellulose (81.3%) and fibres' density of 0.82–0.92 g cm^{-3} [74].

3.1.7   Okra

Okra (*Abelmoschus esculentus* (L.) Moench) is one of the oldest cultivated oligo-purpose crop plant of the Malvaceae family, originated in tropical Africa (Ethiopia and Sudan) and has been now widely spread throughout the tropics and the Mediterranean area. However, it is mainly presented in the Indian and Pakistan region [76, 77]. Okra is mainly cultivated because of its edible green pods and seeds which are used for food. Stems approximately 2 m high are agricultural waste. Mucilage from the stem and roots is used for clarifying sugarcane juice in gur manufacture in India and for sizing paper in China [73]. The stems have also potential for fibre isolation. Fibres can be isolated from stems and branches by similar techniques as used for production of fibres from traditional fibre yielding plants, e.g. dew, cold-, hot- water retting, mechanical and ultrasonic separation, chemical or enzymatic extraction, steam explosion, etc. [78]. Fibres' cross-section has irregular polygonal shape. Fibres' diameter is 40–180 μm and conditioned by plant age and variety and the position of the fibre along the stem [79]. Chemical composition of okra bast fiber is 60–70% cellulose, 15–20% hemicellulose and 5–10% lignin [77, 80]. Alam with co-workers reported the chemical composition as: 67.5% α-cellulose, 15.4% hemicellulose and 7.1% lignin [81]. X–ray diffraction pattern of okra α-cellulose corresponds to typical diffraction pattern of cellulose I. Reflections at $2\theta = 15.6°$, $22.5°$ and $34.0°$ are due to (10-1), (002) and (040) cellulose crystallographic planes, respectively [78]. Based on the fibrillar structure, exhibit fibres good mechanical properties; i.e. tensile strength 24–35 cN/tex and elongation at break 2.5–8.6% [78]. Moisture content is lower than in jute, flax, hemp, etc. and is 4–6% [78]. Besides good qualities fibers show limited resistance to abrasion, poor colour fastness, sensitivity to wear and prone to creasing [76].

Due to fibres' aspect ratio, mechanical properties and low density, okra fibres are a promising material in paper and pulp, furnishing, textile and polymer composites production [79]. Okra fibres were studied as potential reinforcement in biodegradable biopolymer matrix (e.g. poly(lactic acid PLA) for production of biocomposites [82, 83]. In addition to, the extraction of cellulose nanocrystals (CNC) from okra bast fibers has been considered, to produce PVA based nanocomposites [76].

3.1.8   Babylon Willow

Salix babylonica L. (Babylon willow or weeping willow) is a species of the willow family (Salicaceae) native to dry areas of northern China, but cultivated for millennia elsewhere. The Salicaceae family includes the Salix (willow) and Populus (poplar) genera with important species for biomass production [84]. In Europe, Babylon willow was introduced more than 200 years ago. It is a medium- to large-sized deciduous-tree, which grows in hilly and mountainous areas; in particular, it is planted as a decorative tree in parks, promenades and on the banks of rivers. It prefers to grow in moist, loose, fertile and alkaline soils. Tree branches are thin and very long, drifting towards the ground. Leaves are lanceolate, slightly shiny, dark green above, grey-green beneath [85]. Fibres were isolated from the branches of a weeping willow, their intended purpose as a cellulose source for CNC, and some properties were determined (cf. Chap. 6).

In addition, fibres from *Salix miyabeana* Seemen., a species of willow native to northern Japan, were isolated to study the influence of isolation process parameters on fibres' quality. Dilute-acid steam explosion of *Salix miyabeana* was used for fibres' isolation, which were designed for polymer composites' reinforcement. It was found that fibre length, as well as length to diameter aspect ratios, decreased with the severity of the treatment conditions. Fibres' structure was followed and the samples' crystallinity decrease due to reaction temperature was determined [86].

3.1.9 Cassava

Cassava (*Manihot esculenta* Crantz) is a perennial woody shrub in the Euphorbiaceae (spurge family) native to South America, but now cultivated as a root crop largely in many parts of lowland tropics. Some varieties are adapted to dry areas of alkaline soil and others to acid mud banks along rivers. Due to the greatest intraspecific variability Brazil was the most likely area of origin. Archaeological evidence suggests that it was cultivated in Peru 4000 years ago, and in Mexico 2000 years ago. It was introduced to West Africa in the 16th century, and became a major food crop there and in Asia. Cassava starch is used for food and also in the manufacture of adhesives and cosmetics and in sizing textiles, laundering and paper making [10, 73, 87].

Usually 5–10 tubers develop as a swelling on adventitious roots a short distance from stem, old tubers become lignified. Cassava tubers contain starch (70–85% of the root DM), minerals, vitamins, lipids, lignocellulosic fibres but are a pure source of proteins. Modified starches are also produced to improve their functionality for industrial applications so that they can be cooked properly at higher concentrations. Their viscosity is reduced so that their dispersion is improved. Acid-modified starches are clear and stable and may be used for adhesives, gum tapes or bag adhesives. Oxidized starches are used in the paper industry. Cross-linked starches, acetylated starches and cationic starches find their application in starch paper coatings, water-resistant adhesives and coating binders. Biodegradable plastic made from starch is increasingly popular among consumers. The process involves mixing and blending starch with other polymers and stabilizing, gelatinizing and plasticizing agents. Films can be blown that are 40% starch and are biodegradable. Starch can also be incorporated into synthetic polymers to improve plastics for specific uses, such as agricultural plastic mulches and single-use disposable packaging [87].

A purified starch and a solid residue called cassava bagasse (CB) is obtained by industrial exploitation of cassava starch. The total solid residue (dry weight basis) contains 15–50% cellulose fibres [88]. Cassava was applied for cellulose cassava bagasse nanofibres' CBN preparation, therefore, the crystalline structure of cassava fibres was studied. A typical X-Ray Diffraction diagram for cellulose I was observed and the crystallinity index of 43.7% was determined for cassava bagasse fibres [88].

3.1.10 Khimp

Khimp [*Laptadenia pyrotechnica* (Forssk.) Decne.] is a perennial plant, which grows up to 1.8 m, and is belonging to the family Asclepiadaceae. It is a natural desert shrub and grows abundantly in Africa on through desert areas to western India [89]. Fibres are extracted from green stems by crushing and subsequent retting (fibre yield is 8–10%) [90]. From its stems technical fibres consisting of 5–6 ultimate fibres were extracted. Fibres' fineness is 1.01 ± 0.21 tex [1] and the length and diameter of the ultimate fibres 0.67 ± 0.019 cm and 12.9 ± 0.272 µm, respectively [90]. Besides α- cellulose (75.26%) fibres contain lignin (4.93%) and hemicellulose (11.7%) [1]. Cellulose crystallinity index is 60% and is lower than in some conventional fibres, e.g. cotton [1] and the DP of α-cellulose is 904 [90].

Actual today's rural use of fibres is for rope making and animal fodder, however due to fibre properties and availability it has a potential to be used as a replacement for cotton linters and as a raw material for textile, pulp and paper production [90].

3.1.11 Balm-Leaved Archangel

Lamium L. (dead nettle) is a genus comprising 25 species of flowering plants in the family Lamiaceae. It includes both annual and perennial herbaceous plants, native to the temperate and subtropical regions of Asia, Northern Africa and Europe. Representatives of particular species differ in details of flower structure, leaf shape or their arrangement on the stem. Most of the species are characterised by short and toothed lateral lobes of the lower lip of the corolla and a broad and emarginate mid-lobe [91–95].

Balm-leaved archangel (*Lamium orvala* L.) is a herbaceous perennial species, native to central eastern Europe (Austria, Italy, Hungary, Slovenia, Croatia, Ukraine and Moldova) forming a clump of attractive, ovate leaves (up to 15 cm), with spikes of large, velvety purplish pink hooded flowers. The flower is composed of an extremely large upper lip and smaller lower lip.

It is much higher and has larger leaves than other dead nettles. Fine hairs cover leaves' ends. The colour of leaves is usually light green, some leaves have a silver-line along the length. The stem grows from the middle of the roots and is not very branched. It is a non-invasive plant and, unlike other representatives of the family Lamiaceae, it does not proliferate rapidly. However, it is quite resistant. It is a plant with a wide range of soil and condition tolerance, although, it prefers partially shady locations. Most flowers appear in spring [96].

Fibres were isolated from stems and used for CNC preparation (cf. Chap. 6).

3.2 Monocotyledonous Fibre Plants

3.2.1 Agave

The genus *Agave* L. represents more than 200 species within the Asparagoideae subfamily of the Asparagaceae family which are native to the hot and arid regions distributed from the central USA to Panama, Caribbean islands, and to northern tropical areas of South America. Mexico has been reported to have 75% of the *Agave* species. Agave was sacred to the Aztec population of prehispanic Mexico and was named 'Metl'. Charles Linne used *Agave* which is greek for noble and admirable. *Agave* plants are perennial that can grow up to 1.8 m tall, a succulent rosetta is formed by the leaves and it can weigh up to 250 kg in the fresh state. The stem is thick and fibrous and a flower emerges as the stem grows along with the leaf bases and acts as an energy source storage in the form of carbohydrates. Each rosette flowers once and then dies [97–99].

Economic importance includes use by indigenous cultures as a source of fiber which could be utilized for paper and textile production, food, beverages, soap, and medicinals. At present, there is an industrial interest to use the whole plant. Employment of extensive and profitable agricultural and food residues that apparently had no added value is actually the main raw material for the biotechnological industries. *Agave* biotechnology research has been mainly focusing on bio-fuels, beverages, foods, fibers, saponins. However, fibers from *Agave* plants have been little studied: of the more than 200 recorded species, only a handful have been thoroughly analyzed, most notably *A. fourcroydes, A. salmiana, A. sisalana,* and *A. tequilana* [100–104].

The leaves of *Agave sisalana* Perrine are the source of sisal fiber and *A. fourcroydes* Lem. of henequen [99].

A. angustifolia Haw. (synonym *A. vivipara*) is used both to make mezcals and as fiber plant. It has been used to provide food and daily use objects since Aztec times [105]. *A. angustifolia* is additionally commercially cultivated in Mexico, the West Indies and Southern Europe and harvested to produces sisal fibres [97]. Three varieties in particular are cultivated for their fibers: var. *deweyana* (Trel.) in Tamaulipas and Veracruz, var. *letonae* (Taylor) in El Salvador and Guatemala, and var. *nivea* (Trel.) in Guatemala [98].

Fibres of *A. angustifolia* were extracted from leaves and treated additionally by alkali and bleaching treatments. Raw fibres are composed of 67% cellulose, 25.2%

hemicellulose, 6.3% lignin and 2.5% extractives. By spectroscopic methods it was confirmed that hemicellulose and lignin were removed extensively from the extracted fibres by both treatments. The composition of the purified fibre was: 97.3% cellulose, 3.14% hemicellulose and 0.24% lignin. By the X-ray scattering method the index of crystallinity was determined as 59, 69 and 74%, respectively for raw fibres, alkali treated and bleached fibres [106].

Hidalgo-Reyes et al. [103] investigated bagasse fibers from *A. angustifolia*. The fibers were characterized using scanning electron microscopy, differential scanning calorimetry, thermogravimetric analysis, X-ray diffraction and chemical analysis. The tensile strength was also tested using fibers with an uniform length (30 mm). The fibers were light brown in color, with a mean diameter and length of 501 μm and 144 mm, respectively. Scanning electron microscopy images revealed elliptically shaped cells with varying lumen size. Holocellulose content was approximately 82.12%, and total lignin content was approximately 20.69%. Fibers proved to be thermo-stable until 220 °C. Tensile strength, Young's modulus, strain (ε), and ultimate tensile strength were determined via mechanical tests. Due to fibers' thermal properties, low density, and high cellulose content, their future application could be for production of composite materials [103].

In the Mexican food industry blue agave (*Agave tequilana* Weber) is primarily used for tequila alcoholic beverage production since 17th century [107, 108]. The tequila industry has grown substantially in the last 20 years, resulting in an increase in agave production and associated residue (leaves) and bagasse that can be used for second-generation biofuels and textile fiber production [108]. After extraction of sugar for the production of tequila, the residual bagasse, which represent 40% (w/w) of the initial agave biomass, is left. It was demonstrated that industrial by-products of *A. tequilana* have potential for cellulose nanocrystals' CNC extraction. A hydrolysis treatment for degradation of the amorphous region of the cellulose was involved for CNC extraction, producing crystalline nano cellulose based on cellulose type I [109].

3.2.2   Fibre Plants of the Palm Family (Arecaceae)

3.2.2.1   Piassava

Piassava (*Attalea funifera* Mart.) is one of the most conspicuous palm genera in the Neotropics. Ranging from Mexico to Bolivia, Paraguay, Southern Brazil and the Caribbean, it is found in most tropical lowland ecosystems and in the Andes up to 1200–1600 m elevation. Numerous species are massive palms forming dense stands and they are therefore remarkable elements of the landscape, but there are also smaller acaulescent species in both forests and savannas [110, 111].

A stiff highly lignified fibre can be extracted from the leaves of piassava. Over 10,000 tons of this raw material is extracted annually. Piassava fibres that are up to 4 m long and 1.1 mm wide are isolated mechanically once a year for production of ropes, brushes, carpets, etc. About 20% of extracted fibres are waste and these fibres were studied by d'Almeida et al. [112] as potential reinforcement fibres for composite materials.

Fibres have relatively smooth surface topography. On the surface, some star like protrusions were observed constructed of Si-rich particles. The chemical composition of fibres is similar to coir fibres. Both fibres are characterised by a high content of lignin (piassava 48.4% and coir 45%). Cellulose, which is present in 31.6%, shows a X-ray scattering pattern that is typical for cellulose I with the maximum at $2\theta = 24.7°$ [112].

Fibres' mechanical characteristics are typical for lingocellulosic fibres and are comparable to the properties of coir fibres; however, fibres have lower stress at rupture and lower Young's modulus than other lingocellulosic fibres (e.g. jute or sisal). Fibres are a promising candidate as fibre reinforcement for resin—matrix composites [112].

3.2.2.2 Palmyra Palm

The genus *Borassus* L. (Arecaceae) contains five species of massive dioecious fan palms that are distributed across tropical Africa and Asia and New Guinea [113]. Throughout its range, *Borassus* and particularly the Asian palmyra palm (*B. flabellifer* L.) provide a huge variety of products of great significance to local economies. Every part of the palm is utilised, though the most significant product is the sweet sap extracted by tapping the stem or inflorescences. This is fermented into palm wine (arrack) or, the crude sugar (jaggery) is crystallised. The fruits are used as a food product either roasted or raw, and the young, jellylike seeds are also edible. These fruits contain cellulosic semisolid flush, which is reinforced by long coarse and short fine fibres. They are extracted from dried ripened fruits. Coarse fibers are present in the fruit nut edge, whereas the fine fibers were adhered to the shell of the fruit. *Borassus* fine fibres possess superior tensile properties and are an important renewable, natural reinforcement material for composites [113–117]. Because *Borassus* fine fibres (BFF) are inexpensive, abundantly available, eco-friendly, Sudhakara et al. [117] explored fibres extracted from dry ripe fruits for potential use in composites. Composites with PP were prepared successfully by the injection moulding procedure with addition of maleated-polypropylene (MAPP) as a compatibilizer and compared to jute, sisal and coir—polypropilene composites. It was found that, according to the mechanical properties, BFF/PP are comparable to jute/PP and sisal/PP and are superior to coir/PP composites; however, their thermal and sorption characteristics are similar to coir/PP composites [117].

Boopathi, with a group of researchers, treated *Borassus* fruit fibers with 5, 10 and 15% NaOH and explored the effect of alkali treatments on the fibres' properties. Treatment with 5% NaOH yielded significant improvement in tensile properties of fibres [118]. The tensile properties (strength, modulus and elongation) of fibres were improved by 41, 69 and 40% respectively, after 8 h of alkali treatment in 5% NaOH [114].

In addition to, *Borassus flabellifer* petiole fibers were used for preparation of polyester reinforced composites. The neat fibres were extracted and characterised for their tensile and surface characteristics. The composites were fabricated by reinforcing untreated and treated palmyra palm petiole fibre and polyester [116].

3.2.3   Cotton Grass

Cotton grass or cotton sedge (*Eriophorum* L.), is a genus of flowering plants in the sedge family Cyperaceae, native to arctic, subarctic and temperate regions of North America, Asia and Europe found in acid bog habitats on peat and acidic soils, open wetland or moorland. Some species *Eriophorum angustifolium* Honck. and *E. vaginatum* L. are being particularly abundant in Arctic tundra regions. Typical for cotton grass are inflorescences which develop distinctive white seed-heads with wispy, cottony, bright white, red-tinged, or silvery bristles up to 3 centimetres long. Fibres were used for textile production in the 19th century; however, today they are used as an insulating material in replacement of glass wool and for manufacturing waste-water filters. Cotton grass fibre is oleophilic and hydrophobic. It was studied for absorbing oil from the surface of water and it was found that cotton grass is an effective oil sorbent. The absorbing efficiency was over 99% [119].

Today, cotton grass fibre is a constituent of peat fibers which are by-product of peat harvesting industry and are successfully used in the textile industry. Cotton sedge plants arise in bogs where it constantly dies down and then regrows. Peat fibers are obtained from the surface of the bog as a by-product of the energy generation industry. The fibers can be spun into the yarns and used for woven or knitted fabrics [119–121].

Recently, the investigation of Mikucioniene et al. [121] was focused on the thermal and mechanical properties of weft knits made of peat fiber yarns isolated from *Eriophorum vaginatum* sedge stems and cotton fibres and their combinations with other natural fibres, and elastomeric yarn. Knits were designed for clothing. Peat fibres' strength and elongation were compared to cotton and flax yarns. Peat fibres were characterised by high abrasion resistance and their strength and elongation were between properties of cotton and flax yarns [121].

3.2.4   Fibre Plants of the Grass Family (Poaceae)

The true grasses, family Poaceae is a large and nearly ubiquitous family of monocotyledonous flowering plants known as grasses. The groups consist of some 610 genera and about 10,000 species. Grasses are more widespread than any other family of flowering plants and are the most economically important group of monocots, as this family includes the true grains, pasture grasses, sugar cane and bamboo. Different species of the family are widespread and adaptable to different climate conditions from tropical to polar climates. Grasslands cover at least 20% of the earth's surface, although grasses also grow in habitats other than grasslands. Species in this family have been domesticated for staple food crops (grains and sugar, for example), fodder for domesticated animals, biofuel, building materials, paper and ornamental landscaping, among other things [122–124]. Grasses with great potential for fibre industry are represented by bamboo (different taxa of Bambusoideae), sabai grass [*Eulaliopsis binata* (Retz.) C.E. Hubb.], silvergrass (*Miscanthus* Andersson), sugarcane (*Saccharum officinarum* L.), sorghum (*Sorghum bicolor* L. Moench), zoysia grass (*Zoysia* Willd.), perennial ryegrass (*Lolium perenne* L.), Italian ryegrass (*Lolium multiflorum* Lam.), hybrid ryegrasses [*Lolium perenne x multiflorum (Lam.) Parn.*] and tetraploid varieties of perennial and Italian ryegrass. Grasses that were further analysed for renewable fibre utilization and are not discussed above are cat's-tail (*Phleum pratense* L.), cocksfoot (*Dactylis glomerata* L.), fescues (meadow fescue—*Festuca pratensis* Huds.; tall fescue—*F. arundinacea* Schreb.; red fescue—*F. rubra* L.), bromes (*Bromus willdenowii* Kunth) [125, 126]. The length and diameter of ultimate fibres from grasses varies between 0.5 and 3 mm and 10–18 µm, respectively. Fibres from stems are slightly longer than the fibre from the leaves [127].

3.2.4.1 Bamboo Fibres

Bamboo fibres are classified as important plant fibres with great potential for the polymer composite industry. Bamboo fibers possess many excellent properties when used as textile materials such as high tenacity, excellent thermal conductivity, resistance to bacteria, and high water and perspiration adsorption [128].

Bamboos are distributed in tropical and subtropical to mild temperate regions and are evergreen perennial flowering plants in the subfamily Bambusoideae of the grass family Poaceae. Bamboos (subfamily Bambusoideae), are currently comprised of 119 genera and more than 1500 species. However, the taxonomy of bamboo is a difficult domain and the taxonomic demarcation of genera and species at lower ranks are so far not well reserved, because of the infrequent flowering of many species and partly due to extensive genome polyploidization [129, 130]. Bamboos are an agricultural crop that is found worldwide. The greatest diversity is found in Southeast Asia within the Indo-Burmese region, which is also considered to be their place of origin, followed by South America, where bamboo occurs in tropical, subtropical and temperate regions. Fewer bamboo species are found in Africa with exception of Madagascar which is rich in endemic genera and species [130, 131]. Bamboo is extremely fast growing. Some of the fastest-growing bamboo plants, due to a unique rhizome system, can grow about 90 cm within a 24-h period [129]. Bamboo has been used in agriculture, handicrafts, papermaking, furniture and architecture for thousands of years in China [129, 132]. The whole bamboo culm comprises of 50% parenchyma, 40% fibers and 10% conducting tissue (vessels, sieve tubes with companion cells) with some variation according to species [133, 134].

At present, there are two ways to utilise bamboo in the textile industry. One is to isolate fibre bundles from bamboo stems by different treatments. The other method is to spin the regenerated fibre after the bamboo is converted into bamboo pulp. Usually, for the last case, the viscose procedure is involved. In textile production for mass textiles' production the latter method is certainly more prevalent, but isolated fibres also offer many options.

Bamboo fibres are isolated from the bamboo stems according to different processing, i.e. mechanical processing and chemical processing. Splitting of bamboo strips is followed by mechanical or chemical (alkali NaOH hydrolysis) treatment. However, in the mechanical process, the initially crushed bamboo is treated by enzymes and by mechanical comb fibre technology. In the process ultimate fibres are obtained [131].

The width and length of bamboo single fibre are 6–12 μm and 2–3 mm, respectively [132]. Fibres are finer than flax (12–20 μm, 17–20 mm, respectively). Bamboo fibre differs from the other cellulose fibres in morphology. The cross-section of a single bamboo fibre is almost round with a small round lumen. The fibre's surface is without nodes, which are typical for other similar fibres. Bamboo fibre bundles are gummed by 10–20 single fibres [132]. The fineness of a fibre bundle is about 2.58 tex and fibre length 70–90 mm. Its strength is 4.8 cN/dtex [131].

Because of variations in fibre bundles' thicknesses due to the different numbers of elementary fibres in the bundle, there is significant statistical variation in the mechanical properties of fibre bundles. It was found that the average strength of bamboo fibres decreases from 568 to 483 MPa as the mean diameter of the fibre bundles increases from 196.6 to 584.3 μm [135].

Recently, because bamboo fibres represent a potential material within the renewable natural fibres group, it is studied very often and intensively. Bamboo fibres for textile application were investigated by different analytical methods to determine the chemical composition, crystalline structure, molecular and morphology structure [132].

The chemical composition of bamboo fibre is comparable to all other bast fibres. The most important chemical component is cellulose. In addition, fibres contain a high proportion of lignin, and are, therefore, highly woody. Cellulose (60–73%), hemicelluloses (12.5%) and lignin (10.5%) make together about 90% of the total weight of the fibre. The other constituents are protein, fat, pectin, tannins, pigments and ash [131].

Almost axially oriented cellulose fibrils form fibre cell walls (the micro-fibrillar angle is 2–10°). This fibrillar arrangement maximises the longitudinal elastic modulus of the fibres and their lignification increases the transverse rigidity [136].

The Degree of Polymerization of cellulose is lower (DP 891) when compared to other similar fibres [128]. XRD results confirmed that the crystalline structure of the bamboo fibre cellulose belongs to cellulose I; however, there are some differences when the diffraction diagram is compared to flax, cotton and ramie x-ray scattering. The scattering angle of (002) lattice plain is at $2\theta = 22.19°$ with overlapped diffraction peaks of (101) and (10-1) lattice planes at $2\theta = 15.58°$. Analysis of both maxima results in the interplanar dimensions 4.003 and 5.683 nm, respectively. The degree of crystallinity of bamboo fibres is lower than those of flax and cotton, and similar to that of jute, and is 52.45% [128]. In addition to many good qualities, fibres' supramolecular structure is related to some drawbacks, such as bad elasticity, poor wrinkle recovery, harsh handle, low dyeability and itching when worn next to the skin for the fabrics made of them [128].

Bamboo's structural variation, mechanical properties, extraction of fibres, chemical modification, and thermal properties had made this fibres versatile for use in the composite industry as fibre for reinforcement in a variety of matrices [131]. However, bamboo fibre reinforced composites are not utilised fully due to the limited understanding of their mechanical characteristics [137]. Investigated the effects of alkali treatment and elevated temperature on the mechanical properties of bamboo fibre reinforced polyester composites. An improvement in the mechanical properties of the composites was achieved when fibres were treated with NaOH (6%). The bending, tensile, compressive strength and the stiffness of the composite were influenced positively due to the pre-treatment process when compared to the untreated composites [137].

3.2.4.2 Sabai Grass

Sabai grass [*Eulaliopsis binata* (Retz.) C.E. Hubb.] is a perennial plant belonging to the grass family (Poaceae). It is grown in many Asian countries like China, Nepal, India, Pakistan, Thailand, Malaysia and the Philippines. In India, it is being used for paper making since 1870. In some parts in Asia, this species is the second-most important raw material for paper pulp after bamboos, and it is much collected from the wild there. Large-scale cultivation of sabai grass on barren hills and slopes has already been practised and proved to have fast ecological benefits [138–140]. The plant is a good local source of fibre, which is used widely for making sails, ropes and string. Quality of the pulp is similar to that of 'esparto' (produced from *Stipa tenacissima* and *Lygeum spartum)* and is therefore used for high-class book and printing paper. Sabai grass is also popularly used as a construction material for thatches, walls, mats, carpets, roofs and ropes, filler material in plastics and in matrix mud systems [140, 141].

Fibres from sabai grass are long, strong and tough, rich in cellulose and with low lignin content. Sabai grass fibres contain 52.34% of cellulose, 27.2% of hemicelluloses, 16.07% of lignin and 8.12% of moisture [141]. A multicellular fibre consists of several cells with cell size: length 2.4 mm and cell width 9.9 μm [142]. Mechanical properties are: tenacity 3.5–6.9 cN/dtex and elongation 5.0–6.1% [141].

3.2.4.3 Barley

Barley (*Hordeum vulgare* L.) is a member of the grass family (Poaceae). It is one of the first cultivated grains and still one of the major cereal grains grown globally

[143]. It is also a very important crop for the beverage industry, used for the fabrication of beer. Barley husks and straw represent an important agro-industrial waste from the food industry and the possibilities for their utilization were studied by many authors (e.g. [109, 144, 145, 146]).

Sun et al. have studied the chemical composition of differently treated barley straw for extraction of cellulose. According to authors, cellulose yield was ranging between 35.4 and 36.4% (w/w), for H_2O_2 and acetic acid-nitric acid treated straw, respectively. The content of hemicellulose was 2.1–3.2% and lignin 0.5–0.7% (Sun [146]. Barley by-products from food industry were also alkali treated and bleached for isolation of cellulose nanocrystals (CNC) by an acid hydrolysis [109].

3.2.4.4 Ryegrass

The most important representatives of the genus, which are mainly grown for silage, and as a cover crop are perennial ryegrass (*Lolium perenne* L.), Italian ryegrass (*L. multiflorum*) and hybrid ryegrass (*L. perenne* × *L. multiflorum*).

Perennial ryegrass (*Lolium perenne* L.) is one of the most important cool-season grasses in agriculture. It has been cultivated for such a long time by man that its centre of origin is not known. But its original distribution area covers the whole of Europe, temperate Asia and north Africa. It has been introduced to almost all the rest of the world. Europe is generally considered as a centre of variation for this species [147].

Italian ryegrass can be distinguished from *Lolium perenne* (English ryegrass) by the more numerous lemmas in its spikelets (8 or more), which are frequently awned. This plant is grown for silage but became weedy in cereal crops and can be find at ruderal places. Ryegrasses cross-pollinate freely so many types have developed. It is difficult to maintain their genetic purity; consequently, Italian ryegrass is marketed as common ryegrass or domestic ryegrass, and it is often a mixture of annual and perennial species [96, 148]

Properties of ryegrass were studied, i.e. leaf strength, cellulose content and percentage of sclerenchymatic cell walls (percentage of the total leaf cross-section) were studied by Evans [149]. Morphology and mechanical properties of isolated fibres from stems and leaves were studied recently. Ultimate fibres' length is 0.9–1.1 mm in stems and 0.4–0.7 mm in leaves, and linear density of technical fibres consisting of fibre bundles is 12–100 dtex. Mechanical properties of fibre bundles were determined in addition. Fibres' tenacity was 6.2 cN/tex and elongation 1–5% [56]. Besides, ultimate cells dimensions of perennial ryegrass were determined by King et al. [150]. Cell width was 18 and 10 μm for stem and leaf fibres, respectively and fibres' length in the range between 1.32–1.44 mm and 1.02–1.07 mm for stem and leaf fibres, respectively. Differences in individual fibre-cell dimensions are observed across harvest date, grass species and plant part [150].

3.2.4.5 Giant Miscanthus

Silvergrass (*Miscanthus* Andersson), with about 14 species, belongs to Andropogoneae of Poaceae. Various species occur naturally in South-east Asia, the Pacific Islands and tropical Africa [151]. Giant miscanthus (MG) (*Miscanthus* × *giganteus* Greef and Deuter ex Hodkinson and Renvoize) is a large, perennial grass hybrid of *Miscanthus sinensis* and *Miscanthus sacchariflorus*. It has been introduced into Europe and North America as a potential high-yielding energy crop because of its high biomass yield, C4 photosynthesis, stress tolerance, perennial growth and sustainable production. Miscanthus biomass can be used for feedstock, electricity generation or liquid biofuel production, such as ethanol (e.g. [152, 153]). It is also used as a high-quality bedding typically for equine applications. Due to very high C:N ratio it is inhospitable to many microbes thus creating a very clean and antibacterial bedding [154–157]. Influence of the radial stem composition on the thermal behaviour of miscanthus was studied by Chupin et al. [158].

Its chemical composition is 25% hemicellulose, 12% lignin and 50% cellulose [159]. Pulp obtained from *M.* × *giganteus* by alkali treatment and bleaching was used for isolation of cellulose nanocrystals via hydrochloric acid and via sulfuric acid hydrolysis. After alkaline treatment, fibres with the width of 20–50 μm were obtained, however, after bleaching, the width of fibres decreased to 5–10 μm. A significant increase of crystallinity was observed after each treatment. Cellulose crystallinity increased from 48% for MG stalks to 92% for purified pulp (alkaline treatment and bleaching) and 94% for HCl hydrolysed samples [159].

3.2.4.6 Sugarcane Bagasse Fibres

Sugarcane (*Saccharum officinarum* L.) is a tropical grass within Andropogoneae tribe native to South Asia. It is a perennial plant, grows in clumps consisting of a number of strong unbranched stems. From a network of rhizomes under the soil several secondary shoots are grown. It was cultivated in New Guinea about 8000 years ago. Today, over 75% of the world's sugar comes from sugarcane. Sugarcane is grown primarily in the tropics and subtropics. The result of the crossing programmes is that the modem hybrid sugar cane variety incorporates the vigour and hardiness of *S. spontaneum* and *S. sinense* coupled with the higher sugar contents of *S. officinarum* and *S. barberi*. The success of these breeding programmes has contributed to the modem sugar industry being capable of producing 8–17 tonnes sugar per ha [160].

Recently, sugarcane bagasse (SCB) as raw material for industrial applications has become involved increasingly in the process of efficient utilization of agro-industrial residues. After removal of the sugar, the remaining sugarcane bagasse contains about 40–50% cellulose, mainly in crystalline form. 25–35% is hemicelluloses, an amorphous polymer usually composed of xylose, arabinose, galactose, glucose, and mannose. The rest is lignin and some other compounds [161].

3.2.4.7 Sorghum

Sorghum, *Sorghum bicolor* (L.) Moench, which is also known as great millet, belongs to the tribe Andropogonae of the grass family Poaceae. Sorghum is mainly an annual crop, although some have perennial nature in the tropics and can be harvested many times. The greatest variability in both wild and cultivated sorghums is in the north-east Africa. *Sorghum bicolor* (L.) Moench is valued for its grain, stalks and leaves. The crop was domesticated from wild sorghum in Etiopia 5000 years ago. It is one of the World's major cereal crops mostly cultivated by subsistence farmers in the semi-arid tropics of Africa, Asia and Latin America. The Second World War gave impetus to the industrial utilization of sorghum in United States, which is now the world's largest producer of grain sorghum. It has been a subject to intensive selection, and comprises a great array of wild, weedy and annual forms. It has a wide adaptability to harsh growth conditions, such as higher drought, water logging, salinity, and alkalinity tolerance. The cultivated forms are generally classified into five basic cultivar groups. Some species are grown for grain used for flour, malted drinks and beer, while many others are used as fodder plants. Sweet sorghum is any of the many varieties of sorghum whose stalks have a high sugar content. The starch is used in adhesives, for sizing paper and textiles, for gum for stamps and envelopes, and for thickening. The plant bases and stems are used for fuel, thratching, reed walls, fences, baskets and fishtraps ([162, 163]).

Sorghum has been studied for production of cellulose fibres. Fibres from leaves and stems are multi-cellular and have similar cellulose contents. The mechanical properties of isolated fibres were compared to some other cellulose natural fibres. It was observed that the fibres' tenacity and elongation are similar to that of kenaf and cornstalk; however, sorghum fibres' modulus (15 GPa) is similar to the modulus of cornstalk fibres, but higher than that of cotton and cornhusk fibres [164].

Sweet sorghum fibres were studied for preparation of composites with poly (lactic acid). In order to improve fiber-matrix adhesion, the fibres' hydrophilicity was reduced, which is the disadvantage of fibres for preparing composites. The authors suggest four different surface treating agents for surface modification, such as: amino silicone oil, silane coupling agent, emulsified wax and titanate coupling agent [165]. When the process of preparation of the composite is carried out at elevated temperatures, the information is acquired on the thermal behaviour of the fibres. Therefore, Qi and co-workers researched sorghum thermal properties and determined fibres' thermal stability [166]. Composites prepared with whole sorghum stem fragments reinforcing a polyethylene matrix were studied. Different sorghum genotypes were used and the influence of the genotype on composite mechanical properties was followed [167].

3.2.4.8 Zoysia

Zoysia grass (Korean lawngrass; genus *Zoysia* Willd.), belongs to the creeping grasses widespread across Asia and Australia and various islands in the Pacific. The genus *Zoysia* includes 11 species, but only two are considered important turfgrasses species *Z. japonica* Steud. and *Z. tenuifolia* Thiele. The cultivar 'Emerald' (Emerald Zoysia), a hybrid between *Z. japonica* and *Z. tenuifolia*, is particularly popular. It grows well in a wide range of soils, and tolerates heat and drought [168]. A group of Korean researchers isolated nanocrystals from the grass [169]. Cellulose microfibrils, which are embedded in pectin and the hemicellulose matrix, were isolated by alkali treatment and bleaching. Further, samples were hydrolyzed for preparation of CNC. The crystalline structure of isolated grass fibres was studied by X-Ray Diffraction. In the diffraction diagram of native grass, two well-defined peaks were detected at $2\theta = 22.3°$ and $2\theta = 34.5°$. A broader peak at $2\theta = 16.4°$ reflects the scattering from the crystallographic planes (002) and (040) or (10–1) respectively [169].

References

1. Kundu SK, Mojumder P, Bhaduri SK (2005) Physical characteristics of khimp fibre. Indian J Fibre Text Res 30:153–156
2. Reddy N, Yang Y (2005) Biofibers from agricultural byproducts for industrial applications. Trends Biotechnol 23(1):22–27
3. Fiore V, Scalici T, Valenza A (2014) Characterization of a new natural fiber from *Arundo donax* L. as potential reinforcement of polymer composites. Carbohyd Polym 106:77–83
4. Belkhir S, Koubaa A, Khadhri A (2012) Variations in the morphological characteristics of *Stipa tenacissima* fiber: the case of Tunisia. Ind Crops Prod 37(1):200–206
5. Sain M, Panthapulakkal S (2006) Bioprocess preparation of wheat straw fibers and their characterization. Ind Crops Prod 23(1):1–8
6. Prusty JK, Patro SK, Basarkar SS (2016) Concrete using agro-waste as fine aggregate for sustainable built environment—a review. Int J Sustain Built Environ 5(2):312–333
7. Putra A, Abdullah Y, Efendy H (2013) Utilizing sugarcane wasted fibers as a sustainable acoustic absorber. Procedia Eng 53:632–638
8. Navarro-Lisboa R, Herrera C, Zúñiga RN et al (2017) Quinoa proteins (*Chenopodium quinoa* Willd.) fractionated by ultrafiltration using ceramic membranes: the role of pH on physicochemical and conformational properties. Food Bioprod Process 102:20–30
9. Vilcacundo R, Hernández-Ledesma B (2017) Nutritional and biological value of quinoa (*Chenopodium Quinoa* Willd). Curr Opin Food Sci 14:1–6
10. Wickens GE (2001) Economic botany. Kluwer Academic Publishers, Dordrecht
11. van Wyk BE (2005) Food plants of the world—identification, culinary uses and nutritional value. Briza, Pretoria
12. Bhargava A, Shukla S, Ohri D (2006) Chenopodium quinoa—an Indian perspective. Ind Crops Prod 23(1):73–87
13. Sfiligoj-Smole M, Hribernik S, Stana-Kleinschek K et al (2013) Plant fibres for textile and technical applications. In: Grundas S, Stepniewski A (eds) Advances in agrophysical research. InTech, Rijekap, pp 369–397

14. Gasparic P, Urisk Z, Krizanec A et al (2012) Sustainable plant textile fibres. Tekstilec 55 (4):302–313
15. Debnath S (2015) Great potential of stinging nettle for sustainable textile and fashion. In: Gardetti MA, Muthu SS (eds) Handbook of sustainable luxury textiles and fashion. Springer, Berlin
16. Taylor K (2009) Biological Flora of the British Isles: *Urtica dioica* L. J Ecol 97:1436–1458. https://doi.org/10.1111/j.1365-2745.2009.01575.x
17. Bodros E, Baley C (2008) Study of the tensile properties of stinging nettle fibres (*Urtica dioica*). Mater Lett 62:2143–2145
18. Waskow F (1995) Die Renaissance der heimischen Faserpflanzen. Hrsg. Vom Katalyse-institut für Angewandte Umweltforschung. Verlag die Werkstat, Göttingen, pp 93–144
19. Bredemann G (1942) Die Bestimmung des Fasergehaltes bei Massenuntersuchungen von Hanf, Flachs, Fasernesseln und anderen Bastfaserpflanzen. Faserforschung 16:14–39
20. Bredemann G (1959) Die große Brennessel Urtica dioica L. Forschung über ihren Anbau zur Fasergewinnung. Akademieverlag, Berlin
21. Bacci L, di Lonardo S, Albanese L et al (2011) Effect of different extraction methods on fiber quality of nettle (*Urtica dioica* L.). Text Res J 81(8):827–837
22. Baltina A, Lapsa L, Jankauskiene Z et al (2012) Nettle fibers as a potential natural raw material for textile in Latvia. Mater Sci Text Clothing Technol 7:23–27
23. Dreyer J, Edom G (2005) Nettle. In: Franck RR (ed) Bast and other plant fibres. Woodhead Publishing Abington Hall, pp 332–339
24. Bacci L, Baronti S, Predieri S et al (2009) Fiber yield and quality of fiber nettle (*Urtica dioica* L.) cultivated in Italy. Ind Crops Prod 29(2–3):480–484
25. Grundmann E (2007) Versuche zum ökologischen Anbau der Fasernessel. Faserpflanzen aus ökologischem Anbau Tagungsband, Kassel. Schriftenreihe IBDF 20:42–45
26. Pinelli P, Ieri F, Vignolini P, Bacci L, Baronti S, Romani A (2008) Extraction and HPLC analysis of phenolic compounds in leaves, stalks, and textile fibers of *Urtica dioica* L. J Agric Food Chem 56:9127–9132
27. Wurl G, Graf T, Vetter A, Biertrümpfel A (2003) 10 Jahre anbautechnische Versuche zu Fasernesseln (*Urtica dioica* L.) in Thüringen. https://www.tll.de. Accessed 20 Sept 2017
28. Singh R, Bagachi A, Semwal A, Kaur S, Bharadwaj A (2013) Traditional uses, phytochemistry and pharmacology of Morus alba Linn.: a review. J Med Plants Res 7 (9):461–469
29. Binnqüist CL, Quintanar-Isaia, Vander Meeren M (2012) Mexican bark paper: evidence of history of tree species used and their fiber characteristics. Econ Bot 66(2):138–148
30. Peters CM, Rosenthal J, Urbina T (1987) Otomi bark paper in Mexico: commercialization of a pre-hispanic technology. Econ Bot 41(3):423–432
31. Cushman HB, Debo A (eds) (1999) History of the Choctaw, Chickasaw and Natchez Indians. University of Oklahoma Press
32. Dong Z, Ding Z, Zhang S, Zhang Y, Fan H, Yang Y (2017) Natural fibers from the bark of mulberry branches for textile application. Fibres Text Eastern Europe 25, 3(123):20–25
33. Li R, Fei J, Cai Y et al (2009) Cellulose whiskers extracted from mulberry: a novel biomass production. Carbohyd Polym 76(1–2):94–99
34. Qu C, Wang S (2011) Macro-micro structure, antibacterial activity, and physico-mechanical properties of the mulberry bast fibers. Fibers Polym 12(4):471–477
35. Samanta AK, Konar A (2011) Dying of textiles with natural dyes. In: Kumbasar EA (ed) Natural dyes. In Tech, pp 30–31
36. Wang H, Tang Z, Zhou W (2016) A method for dyeing cotton fabric with anthocyanin dyes extracted from mulberry (*Morus rubra*) fruits. Color Technol 132:222–231
37. Neve RA (1991) Hops. Chapman and Hall, London and Van Nostrand Reinhold Publishing Cy, New York
38. Small E (1978) A numerical and nomenclatural analysis of morpho-geographic taxa of Humulus. Syst Bot 3:37–76

39. Jakše J, Kindlhofer K, Javornik B (2001) Assessment of genetic variation and differentiation of hop genotypes by microsatellite and AFLP markers. Genome 44:773–782

40. Seefelder S, Ehrmaier H, Schweizer G et al (2000) Genetic diversity and phylogenetic relationships among accessions of hops, *Humulus lupulus*, as determined by amplified fragment length polymorphism fingerprinting compared with pedigree data. Plant Breed 119:257–263

41. Javornik B, Jake J, Štajner N, Kozjak P (2005) Molecular genetic hop (*Humulus lupulus* L.) research in Slovenia. Acta horticultuae 668:31–34

42. Murakami A, Darby P, Javornik B, Pais MSS, Seigner E, Lutz A, Svoboda P (2006) Molecular phylogeny of wild Hops, *Humulus lupulus* L. Heredity 97:66–74

43. Zanoli P, Zavatti M (2008) Pharmacognostic and pharmacological profile of *Humulus lupulus* L. J Ethnopharmacol 116:383–396

44. DeLyser DY, Kasper WJ (1994) Hopped beer: the case for cultivation. Econ Bot 48(2):166–170

45. Pavlovič M (2014) Hop industry. Quality Management Decision Support Modeling. Verlag Dr. Kovač, Hamburg

46. Reddy N, Yang Y (2009) Properties of natural cellulose fibers from hop stems. Carbohyd Polym 77:898–902

47. Neto WPF, Silvério HA, Dantas NO et al (2013) Extraction and characterization of cellulose nanocrystals from agro-industrial residue—Soy hulls. Ind Crops Prod 42:480–488

48. Leitner J, Hinterstoisser B, Wastyn M et al (2007) Sugar beet cellulose nanofibril-reinforced composites. Cellulose 14:419–425

49. Li ZY, Zhai HM, Zhang Y et al (2012) Cell morphology and chemical characteristics of corn stover fractions. Ind Crops Prod 37(1):130–136

50. Liu Z, Cao Y, Wang Z et al (2015) The utilization of Soybean straw: fibre morphology and chemical characteristics. BioResources 10(2):2266–2280

51. Martelli-Tosi M, da Silva Torricillas M, Martins MA et al (2016) Using commercial enzymes to produce cellulose nanofibers from soybean straw. J Nanomater 2016. https://doi.org/10.1155/2016/8106814

52. Arcioni S, Damiani F, Pezzotti M et al (1990) Alfalfa, lucerne (*Medicago* spp.). In: Bajaj YPS (ed) Biotechnology in agriculture and forestry Vol 10: Legumes and oilseed crops I. Springer, Berlin, pp 197–241

53. Simpson BB, Conner-Ogorzaly M (1986) Economic botany. Plants in our world. McGraw-Hill, New York, pp 497–507

54. Samac DA, Jung HJG, Lamb JAFS (2006) Development of alfalfa (*Medicago sativa* L.) as a feedstock for production of ethanol and other bioproducts. In: Minteer S (ed) Alcoholic fuels. CRC Boca Raton, pp 79–98

55. Ai J, Tschirner U (2010) Fiber length and pulping characteristics of switch-grass, alfalfa stems, hybridpoplar and willow biomasses. Bioresour Technol 101(1):215–221. https://doi.org/10.1016/j.biortech.2009.07.090

56. Sfiligoj-Smole M, Kreze T, Strnad S et al (2005) Characterisation of grass fibres. J Mater Sci 40(20):5349–5353

57. Botha PW, Visser V, Richardson DM et al (2013) Why do French (*Genista monspessulana*) and Spanish brooms (*Spartium junceum*) not sweep across South Africa? S Afr J Bot 86:165

58. Gabriele B, Cerchiara T, Salerno G et al (2010) A new physical-chemical process for the efficient production of cellulose fibers from Spanish broom (*Spartium junceum* L.). Bioresour Technol 101(2):724–729. https://www.ncbi.nlm.nih.gov/pubmed/19734042. https://doi.org/10.1016/j.biortech.2009.08.014

59. Geerts S, Botha PW, Visser V et al (2013) Montpellier broom (*Genista monspessulana*) and Spanish broom (*Spartium junceum*) in South Africa: an assessment of invasiveness and options for management. S Afr J Bot 87:134–145

60. Kovacevic Z, Bischof-Vukusic S, Zimniewska M (2012) Comparison of Spanish broom (*Spartium junceum* L.) and flax (*Linum usitatissimum*) fibre. Text Res J 82(17):1786–1798

61. Sanhueza C, Zalba SM (2012) Experimental control of Spanish broom (*Spartium junceum*) invading natural grasslands. Manage Biol Invasions 3(2):97–104. https://doi.org/10.3391/mbi.2012.3.2.04
62. Swearingen J, Bargeron C (2016) Invasive Plant Atlas of the United States. http://www.invasiveplantatlas.org/
63. Cerchiara T, Chidichimo G, Ragusa MI et al (2010) Characterization and utilization of Spanish Broom (*Spartium junceum* L.) seed oil. Ind Crops Prod 31(2):423–426
64. Angelini LG, Tavarini S, Foschi L (2013) Spanish broom (*Spartium junceum* L.) as new fiber for biocomposites: the effect of crop age and microbial retting on fiber quality. In: Conference papers in materials science 2013, Article ID 274359. https://doi.org/10.1155/2013/274359
65. Katovic A, Katovic D, Antonovic A (2011) Extraction methods of Spanish Broom. Drvna Industrija 62(4):255–261
66. Angelini LG, Lazzeri A, Levita G et al (2000) Ramie (*Boehmeria nivea* (L.) Gaud.) and Spanish Broom (*Spartium junceum* L.) fibres for composite materials: agronomical aspects, morphology and mechanical properties. Ind Crops Prod 11:145–161
67. Kovacevic Z, Krncevic M, Katovic A (2010) Brnistra – zaboravljena tekstilna sirovina. Tekstil 59(9):410–421
68. Cerchiara T, Chidichimo G, Gallucci MC et al (2009) Use of Spanish broom (*Spartium junceum* L.) canvas as a painting support: evaluation of the effects of environmental conditions. J Cult Heritage 10(3):396–402
69. Picuno P (2016) Use of traditional material in farm buildings for a sustainable rural environment. Int J Sustain Built Environ 5(2):451–460
70. Cassano R, Trombino S, Bloise E et al (2007) New broom fiber (*Spartium junceum* L.) derivatives: preparation and characterization. J Agric Food Chem 55(23):9489–9495
71. Cerchiara T, Abruzzo A, Palomino RAÑ et al (2017) Spanish Broom (*Spartium junceum* L.) fibers impregnated with vancomycin-loaded chitosan nanoparticles as new antibacterial wound dressing: Preparation, characterization and antibacterial activity. Eur J Pharm Sci 99:105–112
72. Boynard CA, Monteiro SN, d'Almeida JRM (2003) Aspects of alkali treatment of sponge gourd (*Luffa cylindrica*) fibers on the flexural properties of polyester matrix composites. J Appl Polym Sci 87:1927–1932. https://doi.org/10.1002/app.11522
73. Purseglove JW (1977) Tropical crops dicotyledons. Longman Group, London
74. Siqueira G, Bras J, Dufresne A (2010) Luffa cylindrica as a lignocellulosic source of fibre, microfibrillate cellulose and cellulose nanocrystals. BioResources 5(2):727–740
75. Demir H, Top A, Balköse D, Ülkü S (2008) Dye adsorption behavior of Luffa cylindrica fibers. J Hazard Mater 153:389–394
76. Fortunati E, Puglia D, Monti M et al (2012) Cellulose nanocrystals extracted from okra fibers in PVA nanocomposites. J Appl Polym Sci. https://doi.org/10.1002/app.38524
77. Sathish Kumar D, Tony Eswar, Praveen Kumar A (2013) A review on: *Abelmoschus esculentus* (okra). Int Res J Pharm Appl Sci 3(4):129–132
78. Khan GMA, Yilmaz ND, Yilmaz K (2017) Okra bast fibre as potential reinforcement element of biocomposites: can it be the flax of the future. In: Thakur VK, Thakur MK, Kessler MR (eds) Handbook of composites from renewable materials; functionalization. Wiley, Hoboken
79. Khan GMA, Saheruzzaman M, Razzaque ASM et al (2009) Grafting of acrylonitrile monomer onto bleached okra bast fibre and its textile properties. Indian J Fibre Text Res 34:321–327
80. Khan GMA (2007) Chemical analysis of okra bast fiber (Abelmoschus esculentus) and its physico-chemical properties. J Text Apparel Technol Manage 5(4):1
81. Alam MS, Khan GMA (2007) Grafting of acrylonitrile monomer onto bleached okra bast fibre and its textile properties. Text Apparel Technol Manage 5(4):1

82. Fortunati E, Puglia D, Monti M, et al (2013) Okra (*Abelmoschus esculentus*) fibre based PLA composites: mechanical behaviour and biodegradation. J Polym Environ 21(3). https://doi.org/10.1007/s10924-013-0571-5

83. Santulli C, Sarasini F, Fortunati E et al (2014) Okra fibres as potential reinforcement in biocomposites. In: Hakeem KR, Jawaid M, Rashid U (eds) Biomass and bioenergy; processing and properties. Springer International Publishing Switzerland, pp 176–187. https://doi.org/10.1007/978-3-319-07641-6_11

84. Perdereau AC, Douglas GC, Hodkinson TR et al (2013) High levels of variation in Salix lignocellulose genes revealed using poplar genomic resources. Biotechnol Biofuels 6:114. https://doi.org/10.1186/1754-6834-6-114

85. Zhengyi W, Raven PH, Deyuan H (2004) Flora of China 4. Salicaceae, Science Press (Beijing) & Missouri Botanical Garden (St. Louis). FOC 4:162–274

86. Diop CIK, Lavoie JM, Huneault MA (2015) Structural changes of *Salix miyabeana* cellulose fibres during dilute-acid steam explosion: impact of reaction temperature and retention time. Carbohyd Polym 119:8–17

87. Lebot V (2010) Tropical root and tuber crops. Cassava, sweet potato, yams and aroids. Econ Bot 64(1):86–87.

88. Teixeira EM, Pasquini D, Curvelo AAS (2009) Cassava bagasse cellulose nanofibrils reinforced thermoplastic cassava starch. Carbohyd Polym 78:422–431

89. Masrahi YS (2015) A new species of Leptadenia (Apocynaceae) and two other new records from southwestern Saudi Arabia. Saudi J Biol Sci 22(5):631–636

90. Mojumder P, Mondal SB, Mukhopadhya S et al (2001) Chemical characterization of Khimp fibre (*Leptadenia pyrotechnica*). J Sci Ind Res 60:675–677

91. Atalay Z, Celep F, Bilgili B, Dogan M (2016) Pollen morphology of the genus *Lamium* L. (Lamiaceae) and its systematic implication. Flora 219:68–84

92. Bendiksby M, Brysting AK, Thorbek L, Gussarova G, Ryding O (2011) Molecular phylogeny and taxonomy of the genus *Lamium* L. (Lamiaceae): disentangling origins of presumed allotetraploids. https://www.researchgate.net/publication/266655362_Molecular_phylogeny_and_taxonomy_of_the_genus_Lamium_L_Lamiaceae_Disentangling_origins_of_presumed_allotetraploids. Accessed 20 Sept 2017

93. Govaerts R et al (2014) Lamium in World Checklist of Selected Plant Families. The Board of Trustees of the Royal Botanic Gardens, Kew. Published on the internet. Accessed 2 June 2014

94. Govaerts R, Paton A, Harvey Y, Navarro T (2010) World check list of Lamiaceae and Verbenaceae. Kew, Richmond: The Board of Trustees of the Royal Botanic Gardens. http://www.kew.org/wcsp/lamiaceae/. Accessed 10 Oct 2010

95. Krawczyk K, Głowacka K (2015) Nutlet micromorphology and its taxonomic utility in *Lamium* L. (Lamiaceae). Plant Syst Evol 301:1863–1874. https://doi.org/10.1007/s00606-015-1199-9

96. Martinčič A, Wraber T, Jogan N et al (2007) Mala flora Slovenije: ključ za določevanje praprotnic in semenk. 4. izdaja. Tehniška založba Slovenije, Ljubljana

97. Franck AR (2012) Guide to agave, cinnamomum, corymbia, eucalyptus, pandanus, and sansevieria in the flora of Florida. Phytoneuron 102:1–23

98. Gentry HS (1982) Agaves of Continental North America. University of Arizona Press, Tucson

99. Nava-Cruz NY, Medina-Morales MA, Martinez JL, Rodriguez R, Aguilar CN (2015) Agave biotechnology: an overview. Crit Rev Biotechnol 35:546–559

100. Li Y, Mai YW, Ye L (2000) Sisal fibre and its composites: a review of recent developments. Compos Sci Technol 60(11):2037–2055

101. Piven N, Barredo F, Borges-Argaez C et al (2001) Reproductive biology of henequen (*Agave fourcroydes*) and its wild ancestor *Agave angustifolia* (Agavaceae). I. Gametophyte development. Am J Bot 88:1966–1976. https://doi.org/10.2307/3558424

102. Iñiguez-Covarrubias G, Lange S, Rowell R (2001) Utilization of byproducts from the tequila industry: Part 1: agave bagasse as a raw material for animal feeding and fiberboard production. Bioresour Technol 77:25–32. https://doi.org/10.1016/S0960-8524(00)00137-1.

103. Hidalgo-Reyes M, Caballero-Caballero M, Hernandez-Gomez LH, Urriolagoitia-Calderon G (2015) Chemical and morphological characterization of *Agave angustifolia* bagasse fibres. Bot Sci 93(4):807–817

104. Velazquez-Jimenez LH, Pavlick A, Rangel-Mendez JR (2013) Chemical characterization of raw and treated agave bagasse and its potential as adsorbent of metal cations from water. Ind Crops Prod 43:200–206. https://doi.org/10.1016/j.indcrop.2012.06.049

105. Mohr GM Jr (1999) Blue Agave and its importance in the Tequila Industry. South Ill Univ Carbondale Ethnobotanical Leaflets 3:1–4

106. Rosli N, Ahmad I, Abdullah I (2013) Isolation and characterization of cellulose nanocrystals from *Agave angustifolia* fibre. BioResources 8(2):1893–1908

107. Ruvalcaba-Ruiz D, Rodríguez-Garay B (2002) Aberrant meiotic behavior in Agave tequilana Weber var. Azul. BMC Plant Biol 2(1):10. https://doi.org/10.1186/1471-2229-2-10

108. Valenzuela A (2011) A new agenda for blue agave landraces: food, energy and tequila review. Bioenergy 3:15–24

109. Espino E, Cakir M, Domenek S et al (2014) Isolation and characterization of cellulose nanocrystals from industrial by-products of *Agave tequilana* and barley. Ind Crops Prod 62:552–559

110. Pintaud JC (2008) An overview of the taxonomy of Attalea (Arecaceae). Rev Peruana Biol 15(supl. 1):55–63

111. Pintaud JC, Rodrigez Del Castillo AM, Ferreira EJ, Morales RM, Mejia K (2016) Towards a revision of Attalea in Western Amazonia. Palms 60(2):57–77

112. d'Almeida JRM, Aquino RCMP, Monteiro SN (2006) Tensile mechanical properties, morphological aspects and chemical characterization of piassava (*Attalea funifera*) fibers. Compos A Appl Sci Manuf 37(9):1473–1479

113. Bayton RP (2007) A revision of *Borassus* L. (Arecaceae). Kew Bull 62(4):561–585

114. Reddy KO, Shukla M, Maheswari UC, Rajulu VA (2012) Mechanical and physical characterization of sodium hydroxide treated Borassus fruit fibers. J For Res 23(4):667–674

115. Reddy KO, Maheswari CU, Shukla M (2013) Tensile and structural characterization of alkali treated Borassus fruit finefibers. Compos B Eng 44(1):433–438

116. Srinivasababu N, Kumar JS, Reddy KVK (2014) Manufacturing and characterization of long palmyra palm/Borassus flabellifer petiole fibre reinforced polyester composites. Procedia Technol 14:252–259

117. Sudhakara P, Jagadeesh D, Wang Y et al (2013) Fabrication of Borassus fruit ligno-cellulose fiber/PP composites and comparison with jute, sisal and coir fibers. Carbohyd Polym 98 (1):1002–1010

118. Boopathi L, Sampath PS, Mylsamy K (2012) Investigation of physical, chemical and mechanical properties of raw and alkali treated Borassus fruit fiber. Compos B Eng 43 (8):3044–3052

119. Suni S, Kosunen AL, Hautala M et al (2004) Use of a by-product of peat excavation, cotton grass fibre, as a sorbent for oil-spills. Mar Pollut Bull 49(11–12):916–921

120. Fritz E (1995) The current importance of peat textiles. J Antrophosof Med 12:1–4

121. Mikucioniene D, Cepukone L, Milasiene D (2017) Investigation on mechanical and thermal properties of knits from peat fibers and their combination with other natural fibers. Text Res J. https://www.researchgate.net/publication/316286779_Investigation_on_mechanical_and_thermal_properties_of_knits_from_peat_fibers_and_their_combination_with_other_natural_fibers. Accessed 20 Sep 2017

122. Ahmad F, Khan MA, Ahmad M et al (2009) Taxonomic studies of grasses and their indigenous uses in the salt range area of Pakistan. Afr J Biotech 8(2):231–249

123. Stevens PF (2013) Angiosperm phylogeny website. Version 12. Retrieved from www.mobot.org/mobot/research/APweb/2018

124. Thomas B, Murray BG, Murphy DJ (2017) Encyclopedia of applied plant sciences. Elsevier, Oxford
125. Holmes W (1989) Grass, its Production and Utilization. Blackwell Scientific Publications The British Grassland Society Oxford, London, Edinburgh
126. Petersen A (1981) Die Gräser als Kulturpflanzen und Unkräuter auf Wiese, Weide und Acker Akademie Verlag Berlin
127. Sfiligoj-Smole M, Stana-Kleinschek K, Kreze T et al (2004) Physical properties of grass fibres. Chem Biochem Eng Q 18(1):47–53
128. Liu Y, Hu H (2008) X-ray diffraction study of bamboo fibers treated with NaOH. Fibers Polym 9(6):735–739
129. Hardin IR, Wilson SS, Dhandapani R, Dhende V (2009) An assessment of the validity of claims for "Bamboo" fibers. AATCC Rev 9(10):33–36
130. Nayak L, Mishra SP (2016) Fash Text 3:2. https://doi.org/10.1186/s40691-015-0054-5
131. Khalil HPSA, Bhat IUH, Jawaid M et al (2012) Bamboo fibre reinforced biocomposites: a review. Mater Des 42:353–368
132. Yueping W, Ge W, Haitao C et al (2010) Structures of bamboo fiber for textiles. Text Res J 80(4):334–343. https://doi.org/10.1177/0040517509337633
133. Chaowana P (2013) Bamboo: an alternative raw material for wood and wood-based composites. J Mater Sci Res 2(2):90–102
134. Liese W, Tang TKH, Liese W ed., Kohl M ed. (2015) Properties of the bamboo culm. Tropical forestry, bamboo: the plant and its uses. Switzerland: Springer International Publishing 227–256
135. Wang F, Shao J, Keer LM et al (2015) The effect of elementary fibre variability on bamboo fibre strength. Mater Des 75(15):136–142
136. Wang X, Ren H, Zhang B et al (2012) Cell wall structure and formation of maturing fibres of moso bamboo (*Phyllostachys pubescens*) increase buckling resistance. J R Soc Interface 9 (70):988–996
137. Manalo AC, Wani E, Zukarnain NA et al (2015) Effects of alkali treatment and elevated temperature on the mechanical properties of bamboo fibre–polyester composites. Compos B Eng 80:73–83
138. Duan WJ, Zou DS, Luo JX (2003) The soil and water conservation efficiency of Eulaliopsis binata in the deserted sloping field of purple soil in South China. J Hunan Agric Univ (Nat Sci) 29:204–206
139. Huang Y, Zou DS, Wang H, Yu YL, Luo JX (2003) Ecological benefit of *Eulaliopsis binata* grown in slope wasteland. J Agro-Environ Sci 22:217–220
140. Satpathy AR, Sahu UN (2010) The role of Sabai grass industry in the economic development of Mayurbhanj district of Orissa (India). Int J Appl Agric Res 5(2):221–242
141. Khandual A, Sahu S (2016) Sabai grass: possibility of becoming a potential textile. In: Muthu SS, Gardetti M (eds) Sustainable fibres for fashion industry. Springer Science +Business Media, Singapore, pp 45–60
142. Sahu S, Khandual A, Behera L (2016) Sabai grass fibre: insight into thermal stability, chemical constitution and morphology. Int J Adv Chem Sci Appl 4(4):1–5
143. Newman RK, Newman CW (2008) Barley for food and health: science, technology and products. Wiley
144. Bouasker M, Belayachi N, Hoxha D et al (2014) Physical characterization of natural straw fibers as aggregates for construction materials applications. Materials 7:3034–3048. https://doi.org/10.3390/ma7043034
145. Hyvärinen M, Kärki T (2015) The effects of the substitution of wood fiber with agro-based fiber (barley straw) on the properties of natural fiber/polypropylene composites. In: MATEC web of conferences, vol 30. https://doi.org/10.1051/matecconf/20153001014
146. Sun JX, Xu F, Sun XF et al (2005) Physico-chemical and thermal characterization of cellulose from barley straw. Polym Degrad Stab 88(3):521–531
147. Loos BP (1994) The genus Lolium; Taxonomy and genetic resources. Thesis

148. Quattrocchi U (2006) CRC world dictionary of grasses: common names, scientific names, eponyms, synonyms, and etymology—3 Volume Set. CRC, p 2408
149. Evans PS (1964) A study of leaf strength in four ryegrass varieties. Grasslands Division, Department of Scientific and Industrial Research, Palmerston North
150. King C, McEniry J, Richardson M et al (2014) Characterization for industrial purposes of the fibre anatomy of perennial ryegrass and tall fescue stem and leaf at three stages in the primary growth. Grass Forage Sci 69:64–73. https://doi.org/10.1111/gfs.12032
151. Chen SL, Renvoize SA (2006) Miscanthus Andersson. In: Wu ZY, Raven PH (eds) Flora of China, vol 22. Beijing: Science Press; St. Louis, MO: Missouri Botanical Garden Press, pp 581–583
152. Christian DG, Riche AB, Yates NE (2008) Growth, yield and mineral content of Miscanthus x giganteus grown as a biofuel for 14 successive harvests. Ind Crops Prod 28:320–327
153. Jones MB, Walsh M (2001) Miscanthus for energy and fibre. James & James, London
154. James LEA (2009) Quinoa (Chenopodium quinoa Willd.) Chapter 1, Composition, chemistry, nutritional, and functional properties. Adv Food Nutr Res 58:1–31
155. Sun Q, Lin Q, Yi ZL, Yang ZR, Zhou FS (2010) A taxonomic revision of Miscanthus s.l. (Poaceae) from China. Bot J Linn Soc 164:178–220
156. Vanderghem C, Brostaux Y, Jacquet N, Blecker C, Paquot M (2012) Optimization of formic/ acetic acid delignification of Miscanthus × giganteus for enzymatic hydrolysis using response surface methodology. Ind Crop Prod 35:280–286. https://doi.org/10.1016/j. indcrop.2011.07.014
157. da Silva CG, Grelier S, Pichavant F, Frollini E, Castellan A (2013) Adding value to lignins isolated from sugarcane bagasse and Miscanthus. Ind Crops Prod 42:87–95
158. Chupin L, de Ridder D, Clément-Vidal A et al (2017) Influence of the radial stem composition on the thermal behaviour of miscanthus and sorghum genotypes. Carbohyd Polym 167:12–19. https://doi.org/10.1016/j.carbpol.2017.03.002
159. Cudjoe E, Hunsen M, Xue Z et al (2017) Miscanthus Giganteus: a commercially viable sustainable source of cellulose nanocrystals. Carbohyd Polym 155:230–241. https://doi.org/ 10.1016/j.carbpol.2016.08.049
160. Bakker H (1999) Sugar cane cultivation and management. Springer, New York
161. Sun JX, Sun XF, Zhao H et al (2004) Isolation and characterization of cellulose from sugarcane bagasse. Polym Degrad Stab 84(2):331–339
162. Purseglove JW (1986) Tropical crops monocotyledons. Longman Group, London
163. Reddy PS, Patil JV (eds) (2017) Millets and sorghum biology and genetic improvement. Wiley, West Sussex, United Kingdom
164. Reddy N, Yang Y (2007) Structure and properties of natural cellulose fibers obtained from sorghum leaves and stems. J Agric Food Chem 55(14):5569–5574
165. Zhong J, Li H, Yu J et al (2011) Effects of natural fiber surface modification on mechanical properties of poly(lactic acid) (PLA)/sweet sorghum fiber composites. Polym Plast Technol Eng 50(15):1583–1589
166. Qi C, Yadama V, Guo K et al (2015) Thermal stability evaluation of sweet sorghum fiber and degradation simulation during hot pressing of sweet sorghum–thermoplastic composite panels. Ind Crops Prod 69:335–343
167. Vo LTT, Girones J, Beloli C et al (2017) Processing and properties of sorghum stem fragment-polyethylene composites. Ind Crops Prod 107:386–398. https://doi.org/10.1016/j. indcrop.2017.05.047
168. Wiecko G (2007) Management of tropical turfgrasses. In: Pessarakli M (ed) Handbook of turfgrass management and physiology. CRP Press Taylor & Francis
169. Pandey JK, Lee JW, Chu WS et al (2008) Cellulose nano whiskers from grass of Korea. Macromol Res 16(5):396–398

Chapter 4
Structure and Properties of Non-conventional Cellulose Fibres

Abstract The chemical composition of non-traditional plant fibres depends on their biological origin and age, likewise the extraction method. Chemical composition has a significant impact on the process of isolation of the fibres and their further usability and especially on the fibres' properties.

Keywords Surface properties · Surface morphology · Sorption properties

4.1 Chemical Composition of Non-conventional Cellulose Fibres

The chemical composition of non-traditional plant fibres depends on their biological origin and age, likewise the extraction method. In addition, measurement methods influence the results significantly [1]. Chemical composition has a significant impact on the process of isolation of the fibres and their further usability. Above all, the content of cellulose is very important, while a strong pectine and lignin connection with the tissues surrounding the sclerenchymatic fibre cells, makes it difficult to extract fibres. The large presence of non-cellulosic components extends fibre purification procedures. Table 4.1 provides information on the chemical composition of some fibres, which were extracted by various authors from plants, which have not traditionally been involved in the process of production of fibres. However, lately, these fibres are becoming a potential in sustainable fibre groups. Huge differences in the chemical composition are noticeable. Some are related to the plant source, however the biggest differences are related to the type of the investigated cellulosic material, i.e. isolated, but untreated fibres, treated fibres with removed non-cellulosic compounds (e.g. scoured and bleached fibres), or whole plant fragments containing different cell types. The lowest cellulose content in plant fragments is about 30% or less, but with fibre-isolation process and fibre treatment, it rises to about 97%.

The geometrical characteristics of isolated fibres are collected in Table 4.2.

© The Author(s), under exclusive licence to Springer Nature Switzerland AG 2019   49
M. Sfiligoj Smole et al., *Surface Properties of Non-conventional*
Cellulose Fibres, Biobased Polymers, https://doi.org/10.1007/978-3-030-10407-8_4

Table 4.1 Chemical composition of some non-traditional cellulose fibers

Fibre type	Cellulose (%)	Hemi-cellulose (%)	Lignin (%)	Moisture (%)	References
Banana fibre (raw)	69.9	19.6	5.7	9.8	[21]
Banana fibre (bleached)	96.8	0.2	0.2	9.3	[21]
Pineapple leaf fibre PALF (raw)	75.3	13.3	9.8	9.0	[21]
PALF (bleached)	97.3	0.2	–	8.9	[21]
Jute fibre (raw)	68.3	15.4	10.7	10.1	[21]
Jute (bleached)	97.3	–	–	9.6	[21]
Spanish broom (raw)	44.5 ± 0.2	16.3 ± 0.1	18.5 ± 0.3		[22]
Spanish broom (NaOH and enzyme treated)	91.7 ± 0.1	4.1 ± 0.3	3.2 ± 0.4	9.50	[22]
Bagasse	32–48	19–24	23–32	8.8	[12]
Sugarcane bagasse	43.6	33.5	18.1		[23]
Sisal	43–88	10	3.75–9		[1]
Agave tequilana	45.5 ± 0.1	19.0 ± 3.8	18.7 ± 0.1		[24]
Agave tequilana leaves	54.5	15.2	16.3		[24]
Agave augustifolia	67	25.2	6.3		[25]
Agave augustifolia (alkali treated and bleached)	97.3	3.14	0.24		[25]
Barley husk	31.8 ± 0.1	20.7 ± 2.3	14.9 ± 0.3		[24]
Barley straw	37.7	37.1	15.8	8–12	[12, 24]
Rice straw	59.1	18.4	5.3	6.5	[12, 26]
Wheat straw	43.2 ± 0.2	34.1 ± 1.2	22.0 ± 3.1	10	[12, 27]
Cotton stalks	50.6	28.4	23.1		[26]
Okra	60–70	15–20	5–10	4–6	[28]
	67.5	15.4	7.1		[29]
Soy hulls fibres (raw)	48.2 ± 2.1	24.0 ± 3.0	5.78 ± 1.06		[30]
Soy hulls fibres (purified)	84.6 ± 4.0	11.2 ± 4.0	3.67 ± 0.3		[30]
Khimp	75.26	11.7	4.93		[31]
Soy bean straw—raw	35	17	21		[32, 33]
Corn stover	38–40	28	7–21		[12]
Sorghum stalks	27	25	11	8–12	[12]
Luffa cylindrica	65.5 ± 0.5	17.5 ± 0.5	15.2 ± 1.0		[34]
Mengkuang leaves (*Pandanus tectorius*)	37.3 ± 0.6	34.4 ± 0.2	24 ± 0.8		[35]
Mengkuang leaves (alkali treated)	57.5 ± 0.8	15.5 ± 0.1	22.6 ± 0.2		[35]
Mengkuang leaves (bleached)	81.6 ± 0.6	15.9 ± 0.6	0.8 ± 0.1		[35]

(continued)

Table 4.1 (continued)

Fibre type	Cellulose (%)	Hemi-cellulose (%)	Lignin (%)	Moisture (%)	References
Napier grass fibre (untreated)	46.58	34.14	22.25		[36]
Napier grass fibre (2%NaOH treated)	55.28	25.24	21.75		[36]
Napier grass fibre (2% acetic acid treated)	56.46	20.55	20.93		[36]
Sabai grass	52.34	27.2	16.07	8.12	[37]
Zostera marina	57	38	5		[38]
Nettle	54	10	9.4	11.5–12.17	[39]
Lucerne	30	12	17		[40]
Arundo	43.2	20.5	17.2		[41]
Artichoke	75.3		4.3		[41]
Ferula	53.3	8.5	1.4		[41]
Althaea	44.6	13.5	2.7		Fiore [41]

Table 4.2 Fibres' length and diameter

Fibre type	Ultimate fibre length (mm)	Fibre diameter (μm)	References
Agave tequilana	0.315	47.7	[24]
Barley husks	0.429	46.3	[24]
Corn husk	0.5–1.5	10–20	[12]
PALF	3–9	20–80	[12]
Bagasse	0.8–2.8	10–34	[12]
Wheat straw	0.4–3.2	8–34	[12]
Khimp	6.7 ± 0.019	12.9 ± 0.272	[42]
Rice straw	0.4–3.4	4–16	[12]
Sorghum stalks	0.8–1.2	30–80	[12]
Barley straw	0.7–3.1	7–24	[12]
Agave augustifolia		7–12	[25]
Soybean straw	0.46	24.2	[43]
Musa maturate rachis		10–20	[44]
Giant reed	1.18	15	[45]
Bamboo	2–3	6–12	[46]
Sabai grass	2.4	9.9	[47]
Spanish broom	25–27	7–10	[22]
Nettle	43–58	19–47	[48]

4.2 Surface Properties

4.2.1 Surface Morphology

Plant fibrous cells have a complex structure on the surface. Elementary fibres are, in the natural state, assembled into bundles of 10–40 fibres, maintained together by polysaccharides (composed mainly of pectin) [2]. The boundaries between the various cells in the bundle are clearly visible in the micrographs (cf. Figs. 4.1 and 4.2).

The non-uniform geometrical characteristics, i.e. a geometrically irregular cross-section shape and non-constant transverse dimension, are typical for stem and leaves' fibres [2].

Like many other plant tissues, the cell walls of lignocellulose fibres are a multi-layered structure composed of middle lamella, a primary, a secondary wall and a cell lumen in the middle. The complexity of the fibre structure results in an inhomogeneous surface with many more or less expressed surface structures. Noticeable are morphological effects, especially the effect of roughness and porosity.

Fig. 4.1 SEM image of hemp and rye-grass fibres

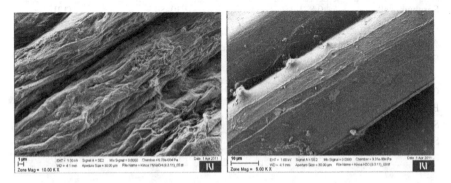

Fig. 4.2 SEM image of quinoa fibres isolated by chemical treatment (NaOH) and water retted

Lignocellulose fibre surface cannot be compared with the appearance of the smooth surface of chemical fibres with no surface characteristics. In addition, the applied fibre separation processes affect the fibre surface morphology, which also influences the specific fibre surface area and, thereby, several fibre properties [3].

Furthermore, a number of chemical or physical treatments affect the surface structure of fibres. The impact of the treatment on surface morphology is depending on the treatment conditions, i.e. if they are excessive, the surface changes are more pronounced and can be reflected as damages. Alkali treatment removes surface impurities and additionally some of hemicellulose and lignin components depending on the alkali concentration and treatment time. In parallel, the activity of the alkali is reflected in morphological and topological changes. Additional impacts are elevated temperature and ultrasound which are used to influence the treatment efficiency, however such conditions can lead to fibre damages and defects [4]. Finally, such processing may lead to fibrillation, which means the cleavage of individual fibrils, which significantly modifies the surface topography [4].

Figures 4.1 and 4.2 show surface morphologies of some plant fibres with potential for different applications. SEM micrographs show uneven fibre surface. Non-cellulosic materials are covering fibre surface and thereby give a rough surface appearance. Fibres are not separated into ultimate fibres but are bundles of fibrous cells, which is clearly visible in the images.

Surface differences due to the fibres isolation technique are demonstrated clearly on the longitudinal views of quinoa fibres in Fig. 4.2. Fibres have been isolated from the stalks by water treatment and in alkaline medium, respectively. The latter shows a non-smooth corrugated surface that originates from the action of sodium hydroxide on the fibre. A rougher surface with ridges is noticeable.

Surface morphology plays an important role in composites. The adhesion between the reinforcing fibres and the polymer matrix is also conditioned by surface topography and, consequently, it influences the final mechanical properties of the composites, since the stress transfer between the matrix and fibres determines reinforcement efficiency [5].

The surface of the fibre can also be changed depending on the moisture content in the fibre, as it is known that the lignocellulose fibres are very hydrophilic, which causes their swelling. Fibres' poor dimensional stability due to swelling is a dis-advantage of these fibres. ζ-potential reflects clearly the differences in the degree of hydrophilicity of the natural fibres and, therefore, it is an appropriate method for fibres' wet state definition [3].

Often in the preparation of composites, non-polar matrix materials are combined with polar reinforcing fibres such as natural plant fibres. The inherent polar and hydrophilic nature of natural fibres and the non-polar characteristics of matrix materials result in difficulties, e.g. poor stress transfer efficiencies [3]. Therefore, the knowledge of the properties of both, bulk and surfaces of materials that are com-pounded in a composite material is essential for designing a composite.

4.2.2 Sorption Properties

The tendency of a liquid for making a close contact with a surface is defined as wettability [6]. Reactivity, accessibility and adsorption properties of fibres are extremely important fibre properties for their application and treatment processes. Sorption properties are influenced by the chemical composition and the complex porous and fibrillar morphology, i.e. the size of crystallites and amorphous regions, amorphous and crystalline orientation, size and shape of the voids and the number of interfibrillar lateral tie molecules [7–9], thus chemically heterogeneous, rough and porous plant fibres surfaces are strongly influencing the wetting phenomena of fibres. In addition too, the purity of the fibre, which depends on the level of removal of non-cellulosic hydrophobic substances in the primary cell wall of lignocellulose fibres, also has a tremendous impact on the sorption properties of the fibre. The main components of lignocellulose fibres, i.e. cellulose, hemicellulose and lignin, are significantly different in their wetting-ability.

In cellulose, which is generally the dominant component in the chemical composition of the lignocellulosic fibres, are the most notable functional groups –OH groups, which form numerous intra- and inter- molecular hydrogen bondings, on which the stability of the regular crystal structure is based. Despite the fact that there are so many –OH groups inaccessible, they are still the most important functional groups in the adhesion and wettability processes [6].

Regarding wettability, the presence of hemicelluloses in natural plant fibres is very important. Hemicelluloses, because of its irregular nature, are present mainly in the irregular non-crystalline structure, and thus with their side groups (e.g. carboxylic acid –COOH group, etc.) they are providing better accessibility for the interactions. Furthermore, it has been shown that swelling of fibres with high hemicelluloses content is greater than the others [6].

Lignin is one of the most abundant natural polymers and is a complex irregularly structured highly aromatic polymer that forms support tissues of plants [10]. Lignins are cross-linked phenolic polymers with monomeric units–guaiacyl, -syringyl, and –p-ydroxyphenyl randomly bound by ester, ether, –C–C– and other connections [6]. Native lignin is much more hydrophobic in comparison to cellulose and hemicellulose.

In order to characterise fibres wettability, interaction with water is often employed, which is able to destroy weaker hydrogen bonds in cellulose, but cannot penetrate into the regions of high order [11].

Whereas the sorption properties are one of the most important functional properties of fibres, it is very important that, fiber sorption properties are considered when designing a product. This is particularly important, especially in multi-component materials such as composites. When a fibre with high hydrophilicity, like cellulose, is planned as the reinforcing phase in the hydrophobic matrix, the fibres' moisture sorption and consequent swelling could affect adhesion between phases of the composite significantly and, thus, the deterioration of properties, especially mechanical properties.

Table 4.3 Moisture sorption properties of fibres isolated from nettle, hemp, sorghum, ryegrass, quinoa and corn-stalks

Fibre type	Moisture content M_s (%)
Nettle	11.19
Sorghum	11.15
Cornstalk	10.66
Ryegrass	10.3
Hemp	12.06
Quinoa	13.39

Moisture sorption was determined on water retted and manually isolated fibres of nettle, ryegrass, sorghum, cornstalk, hemp and quinoa. After isolation, fibres were, additionally, milled in a grinder and weighed. The method according to the Standard DIN 54 351 was used to determine the moisture uptake of different fibres from non-traditional plant sources. After exposing fibres to Standard atmosphere (ISO 139:2005 Textiles—Standard atmospheres for conditioning and testing) T = 20 ± 2 °C, RH = 65% ± 2% t = 24 h, moisture sorption was calculated as the mass % of absolute dry material according to Eq. 4.1. The results are collected in Table 4.3.

$$M_s = (m_1 - m_0) \cdot 100/m_0 \qquad (4.1)$$

M_s moisture sorption (%)
m_0 mass of absolute dry material
m_1 mass of sample after exposure to 65% RH

Fibres from nettle, ryegrass, sorghum, cornstalk, hemp and quinoa indicate high water sorption ability that is typical for all cellulose fibres; e.g. the moisture content of flax is 8–10% [2], sorghum stalks 8–12%, barley 8–12%, rice straw 6.5% [12], Spanish broom 9.50% [13] and hemp is 12% [14]. For Spanish broom fibres the relation between relative humidity and fibres' moisture content was measured. For relative humidity 20, 65 and 95%, respectively fibres moisture content of 2.62, 9.50 and 12,82%, respectively were determined [13]. By lignocellulose fibres, removing non-cellulosic typically more hydrophobic substances, can influence the quality of water sorption properties. Therefore, techniques of fibres' isolation and additional pre-treatment processes have an important role in the hydrophilic-hydrophobic behaviour of lignocellulosic fibres.

Besides the moisture content measuring method, other methods are also available for determination of fibres' sorption properties, e.g. water retention power, which is determined according to the Standard DIN 53 814. The method is based on the water quantity, which can be absorbed and retained under strictly controlled conditions. It is expressed as the ratio between the mass of water retained in the fibre after soaking (2 h) and centrifuging (20 min), and the mass of absolute dry sample [9].

An additional method for water sorption properties quantification is determining the contact angle by its direct measuring (optical method) and by tensiometry.

Contact angle is a very important quality parameter of solid surfaces, as it is the base for understanding and interpretation of surface properties such as adhesion, wetting and surface free energy.

Tensiometry is a method for measuring the reactivity (adsorption character) of a fibre in a polar environment. The method measures an interaction force between a liquid and a solid as the liquid is brought into contact with the solid [15]. The method measures fibres' mass during the adsorption of liquids. The method is described in detail in [16, 17]. The time dependence of the samples' square mass is recorded graphically and capillary velocity is determined from the slope of the constant initial part of the curve. The modified Washburn equation is used [16] for calculation of the contact angle.

In order to determine the sorption abilities, the capillary velocities of some non-traditional fibers were measured on Tensiometer Krüss K12, GmbH, (Germany) and used for calculations of contact angles according to the modified Washburn equation. The capillary constant was obtained by wetting the samples by n–heptan and subsequently the wetting of water on samples was performed. Figure 4.3 shows the diagrams of the rate of water absorbtion onto fibers isolated from nettle and quinoa. The results of contact angles are presented in Table 4.4.

Eight parallel samples of two types of untreated lignocellulose fibres; i.e. nettle and quinoa, were measured and the average contact angles for both fibres are above 80°. The results are in good agreement with the results obtained by other researchers for similar plant fibres. Contact angles obtained for raw cotton fibres (95.3° ± 19.4°) and grey plain weave cotton fabric (93.9° ± 3.3°) decreased after alkaline scouring of fabric (41.3° ± 5.4°) [18]. The contact angle of single bamboo fibres determined by the optical method was 62°and it was higher than ramie (55°) but lower than that of jute (64°) after the same chemical treatment [19]. Orue with

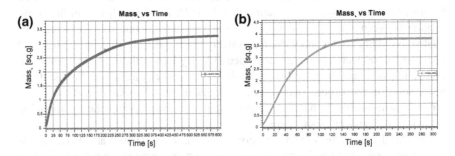

Fig. 4.3 Sorption diagrams of fibres isolated from nettle (**a**) and quinoa (**b**)

Table 4.4 Contact angles of fibres isolated from nettle and quinoa

Fibre type	Contact angle (°)
Nettle	84.8
Quinoa	82.9

co-workers reported about sisal fibers, which were surface modified by sodium hydroxide and 3-(2-aminoethyl amino) propyltrimethoxy silane. Contact angle values and the absorption of a water drop in fibres was measured on compressed discs of sisal fibers. For the untreated sisal fibres, NaOH treated, silane treated and NaOH and silane treated fibres contact angles 68.2° ± 0.3°, 75.4° ± 0.7°, 88.2° ± 2.1°, 94.8° ± 1.7°, respectively were obtained [20].

References

1. Li Y, Mai YW, Ye L (2000) Sisal fibre and its composites: a review of recent developments. Compos Sci Technol 60(11):2037–2055
2. Baley C, Busnel F, Grohens Y et al (2006) Influence of chemical treatments on surface properties and adhesion of flax fibre–polyester resin. Compos: Part A 37:1626–1637
3. Bismarck A, Aranberri-Askargorta I, Springer J et al (2002) Surface characterization of flax, hemp and cellulose fibers; surface properties and the water uptake behavior. Polym Compos 23(5):872–895
4. Cho D, Kim HJ, Drzal LT (2014) Surface treatment and characterization of natural fibers: effects on the properties of biocomposites. In: Thomas S, Joseph K, Malhotra SK et al (eds) Polymer composites, biocomposites, vol 3, 1st edn. Willley, Weinheim
5. Han SO, Choi HY (2010) Morphology and surface properties of natural fibre treated with electron beam In: Mendez-Vilas A, Diaz J (eds) Microscopy: science, technology, applications and education, vol 3, pp 1880–1887
6. Hubbe MA, Gardner DJ, Shen W (2015) Contact angles and wettability of cellulosic surfaces: a review of proposed mechanisms and test strategies. BioResources 10(4):8657–8749
7. Schurz J (1994) Was ist neu an den neuen Fasern der Gattung Lyocell? Lenzinger Berichte 74:37–40
8. Schurz J, Lenz J (1994) Investigations on the structure of regenerated cellulose fibers. Macromol Symp 83(1):273–289
9. Kreze T, Stana-Kleinschek K, Ribitsch V (2001) The sorption behaviour of cellulose fibres. Lenzinger Berichte 80:28–33
10. Maximova N, Österberg M, Laine J et al (2004) The wetting properties and morphology of lignin adsorbed on cellulose fibres and mica. Colloids Surf A: Physicochem Eng Aspects 239 (1–3):65–75. https://doi.org/10.1016/j.colsurfa.2004.01.015
11. Peter M, Rouette HK (1989) Grundlagen der Textilveredlung, 13th edn. Deutscher Fachverlag, Frankfurt
12. Reddy N, Yang Y (2005) Biofibers from agricultural byproducts for industrial applications. Trends Biotechnol 23(1):22–27
13. Cerchiara T, Chidichimo G, Rondi G et al (2014) Chemical composition, morphology and tensile properties of Spanish broom (*Spartium junceum* L.) fibres in comparison with flax (*Linum usitatissimum* L.). Fibres Text Eastern Europe (22)2: 25–28
14. Kymäläinen HR, Pasila A (2000) Equilibrium moisture content of flax/linseed and fibre hemp straw fractions. Agric Food Sci Finland 9:259–268
15. Cao Y, Chan F, Chui YH et al (2012) Characterization of falx fibres modified by alkaline, enzyme and steam-heat treatments. BioResources 7(3):4109–4121
16. Persin Z, Stana-Kleinschek K, Kreze T (2002) Hydrophilic/hydrophobic characteristics of different cellulose fibres monitored by tensiometry. Croat Chem Acta 75(1):271–280
17. Persin Z, Stana-Kleinschek K, Sfiligoj-Smole M et al (2004) Determining the surface free energy of cellulose materials with the powder contact angle method. Text Res J 74(1):55–62. https://doi.org/10.1177/004051750407400110

18. Hsieh Y-L, Thompson J, Miller A (1996) Water wetting and retention of cotton assemblies as affected by alkaline and bleaching treatments. Text Res J 66(7):456–464
19. Chen H, Cheng H, Jiang Z et al (2013) Contact angles of single bamboo fibers measured in different environments and compared with other plant fibers and bamboo strips. BioResources 8(2):2827–2838
20. Orue A, Eceiza A, Peña-Rodriguez C et al (2016) Water uptake behavior and young modulus prediction of composites based on treated sisal fibers and poly(lactic acid). Materials 66 (7):456–464. https://doi.org/10.3390/ma9050400
21. Abraham E, Deepa B, Pothan LA et al (2011) Extraction of nanocellulose fibrils from lignocellulosic fibres: a novel approach. Carbohyd Polym 86:1468–1475
22. Gabriele B, Cerchiara T, Salerno G et al (2010) A new physical-chemical process for the efficient production of cellulose fibers from Spanish broom (*Spartium junceum* L.). Bioresour Technol 101(2): 724–729. https://doi.org/10.1016/j.biortech.2009.08.014, https://www.ncbi.nlm.nih.gov/pubmed/19734042
23. Sun JX, Sun XF, Zhao H et al (2004) Isolation and characterization of cellulose from sugarcane bagasse. Polym Degrad Stab 84(2):331–339
24. Espino E, Cakir M, Domenek S et al (2014) Isolation and characterization of cellulose nanocrystals from industrial by-products of Agave tequilana and barley. Ind Crops Prod 62:552–559
25. Rosli N, Ahmad I, Abdullah I (2013) Isolation and characterization of cellulose nanocrystals from *Agave angustifolia* fibre. BioResources 8(2):1893–1908
26. Habibi Y, El-Zawawy WK, Ibrahim MM et al (2008) Processing and characterization of reinforced polyethylene composites made with lignocellulosic fibers from Egyptian agroindustrial residues. Compos Sci Technol 68(7–8):1877–1885. https://doi.org/10.1016/j.compscitech.2008.01.008
27. Alemdar A, Sain M (2008) Biocomposites from wheat straw nanofibers: morphology, thermal and mechanical properties. Compos Sci Technol 68(2):557–565
28. Khan GMA, Saheruzzaman M, Razzaque ASM et al (2009) Grafting of acrylonitrile monomer onto bleached okra bast fibre and its textile properties. Indian J Fibre Text Res 34:321–327
29. Alam MS, Khan GMA (2007) Grafting of acrylonitrile monomer onto bleached okra bast fibre and its textile properties. Text Apparel Technol Manage 5(4):1
30. Neto WPF, Silvério HA, Dantas NO et al (2013) Extraction and characterization of cellulose nanocrystals from agro-industrial residue—Soy hulls. Ind Crops Prod 42:480–488
31. Kundu SK, Mojumder P, Bhaduri SK (2005) Physical characteristics of khimp fibre. Indian J Fibre Text Res 30:153–156
32. Leitner J, Hinterstoisser B, Wastyn M et al (2007) Sugar beet cellulose nanofibril-reinforced composites. Cellulose 14:419–425
33. Martelli-Tosi M, da Silva Torricillas M, Martins MA et al (2016) Using commercial enzymes to produce cellulose nanofibers from soybean straw. J Nanomat. https://doi.org/10.1155/2016/8106814
34. Siqueira G, Bras J, Dufresne A (2010) Luffa cylindrica as a lignocellulosic source of fibre, microfibrillate cellulose and cellulose nanocrystals. BioResources 5(2):727–740
35. Sheltami RM, Abdullah I, Ahmad I et al (2012) Extraction of cellulose nanocrystals from mengkuang leaves (*Pandanus tectorius*). Carbohyd Polym 88(2):772–779
36. Kamarullah SH, Mydin MM, Atikah WS et al (2015) Surface morphology and chemical composition of napier grass fibers. Malays J Anal Sci 19(4):889–895
37. Khandual A, Sahu S (2016) Sabai grass: possibility of becoming a potential textile. In: Muthu SS, Gardetti M (ed) Sustainable fibres for fashion industry. Springer Science+Business Media, Singapore, pp 45–60
38. Davies P, Morvan C, Sire O, Baley C (2007) Structure and properties of fibres from sea-grass (*Zostera marina*). J Mater Sci 42:4850–4857
39. Dreyer J, Edom G (2005) Nettle. In: Franck RR (ed) Bast and other plant fibres. Woodhead Publishing, Abington Hall, pp 332–339

40. Samac DA, Jung HJG, Lamb JAFS (2006) Development of alfalfa (*Medicago sativa* L.) as a feedstock for production of ethanol and other bioproducts. In: Minteer S (ed) Alcoholic fuels. CRC Boca Raton, pp 79–98

41. Fiore V, Scalici T, Valenza A (2014) Characterization of a new natural fiber from *Arundo donax* L. as potential reinforcement of polymer composites. Carbohyd Polym 106:77–83

42. Mojumder P, Mondal SB, Mukhopadhya S et al (2001) Chemical characterization of khimp fibre (*Leptadenia pyrotechnica*). J Sci Ind Res 60:675–677

43. Liu Z, Cao Y, Wang Z et al (2015) The utilization of soybean straw: fibre morphology and chemical characteristics. BioResources 10(2):2266–2280

44. Gañán P, Zuluaga R, Cruz J et al (2008) Elucidation of the fibrous structure of Musaceae maturate rachis. Cellulose 15:131–139

45. Marques G, Rencoret J, Gutiérrez A et al (2010) Evaluation of the chemical composition of different non-woody plant fibers used for pulp and paper manufacturing. Open Agric J 4:93–101

46. Yueping W, Ge W, Haitao C et al (2010) Structures of bamboo fiber for textiles. Text Res J 80(4):334–343. https://doi.org/10.1177/0040517509337633

47. Sahu S, Khandual A, Behera L (2016) Sabai grass fibre: insight into thermal stability, chemical constitution and morphology. Int J Adv Chem Sci Appl 4(4):1–5

48. Bacci L, Baronti S, Predieri S et al (2009) Fiber yield and quality of fiber nettle (*Urtica dioica* L.) cultivated in Italy. Ind Crops Prod 29(2–3):480–484

Chapter 5
Cellulose Nanofibres

Abstract Nanocrystals (CNCs) and cellulose nanofibrils (CNFs) are the two main types of nanocellulose. Recently several non-wood lignocellulosic sources are studied for preparation of cellulose nanomateriales and also many non-traditional sources of cellulose are involved in research, such as soybean stock, swede root, sugar beet pulp, potato tubers, wheat straw, chardonnay grape skins, cassava bagasse, cotton stalk, bamboo, agave T, barley husks, giant miscanthus, switch grass, water hyacinth, mulberry bark, loofah, zoysia grass, etc.

Keywords Nanocellulose · Cellulose nanofibres from non-traditional cellulose sources

Nanocellulose and its composites have been studied intensively recently. Related to the nanodimensions of nanocellulose, a high surface area is formed and, hence, the powerful interaction of these celluloses are observed with surrounding species, such as water, organic and polymeric compounds, nanoparticles, and living cells [1]. Nanocellulose has a great potential for use in nanocomposites as reinforcement, and has attracted great interest recently. It has exceptional mechanical properties and, because of the small dimensions of nanofibers, being less than one-tenth of the visible light wavelength, cellulose nanofibres are an ideal reinforcement of transparent resins, since they are free from light scattering [2]. Nanocellulose based nanocomposites generally exhibit significant improvements in thermal, mechanical, and barrier properties compared to the neat polymer or conventional composites [3]. The advantage of nanocellulose is its low density (about $1.566 \ \mathrm{g \ cm^{-3}}$), non-abrasive nature, non-toxic character, biocompatibility and biodegradability and its origin from renewable natural sources, which are very abundant and, therefore, low-costing [3].

The morphology and properties of nanocellulose, which depend mainly on the source of the original cellulose and the extraction process, influence their performance as a reinforcing agent [3]. It was observed that significant changes in mechanical properties, especially maximum load and tensile strength, are obtained for nanocomposites with nanocellulose from different cellulosic sources [4].

© The Author(s), under exclusive licence to Springer Nature Switzerland AG 2019
M. Sfiligoj Smole et al., *Surface Properties of Non-conventional*
Cellulose Fibres, Biobased Polymers, https://doi.org/10.1007/978-3-030-10407-8_5

It was confirmed that it is a very versatile material, having a wide range of medical applications, including cardiovascular implants, scaffolds for tissue engineering, repair of articular cartilage, vascular grafts, urethral catheters, mammary prostheses, adhesion barriers, and artificial skin [5]. Implants are produced from bioresorbable and/or biodegradable materials. In addition, nanocellulose is also suitable for use in food products as thickeners or stabilizers, in paintings, cosmetics, automotive materials, packaging, gas barrier films and optically transparent functional materials, etc.

Nanocellulose, nanocrystals, whiskers, rods, nanofibrils, or nanofibres, are, when the cellulose fibre or crystal has at least one dimension within the nanometer size range [6, 7]. The dimensions of nanofibrils are usually around 20–30 nm in diameter with a length of a few μm. Nanocrystals are much smaller. Their length is about 200 nm and diameter about 3–5 nm [8].

Two main types of nanocellulose are distinguished: (i) The one obtained by acid treatment, referred to as nanocrystals (CNCs) and (ii) The ones produced mainly by mechanical disintegration called cellulose nanofibrils (CNFs) and cellulose microfibrils (CMFs) [9, 10].

Different cellulose sources give different characteristics of CNCs and CNFs and also different aspect ratios (L/d, where L is the length and d is the diameter). A typical aspect ratio for CNC ranges from 1 to 100 and for CNF from 15–100 [7].

Cellulose nanofibrils CNFs are separated fibrils after the interfibrillare hydrogen bonds break due to different treatments; however, the fibrillar structure is preserved composed from periodical arrangement of crystalline and amorphous domains. Fibres are obtained with micrometre size in length and nano-metre size in width [7].

CNC particles are rod-like with widths and lengths of 5–70 nm and 100 nm—several micro-meters, respectively. These are cellulose, highly crystalline particles (index of crystallinity 54–88%) [11]. Wood CNC has a width between 3 and 5 nm and length between 100 and 200 nm [11]. Nanocrystals obtained from cotton are 100–300 nm in length, and 8–10 nm in diameter, while those of the tunicate nano-crystals range from 100 nm to a few micrometres in length and 10–20 nm in diameter [12].

The use of nanocrystals obtained from different cellulose sources in the field of nano-composites has attracted considerable attention since it was first reported a few decades ago [13]. The applicability of CNC is very versatile, e.g. for coatings, pharmaceuticals, reinforced plastics, textiles, energy storage, sound insulators, acoustic membranes, recyclable oil absorbers, paper applications, packaging, air filtration, etc. Its potential is also in applications that require a lower weight material, such as in spacecrafts, airplanes, automobiles, etc. [14].

Nanocrystalline cellulose is constructed mainly from crystalline cellulose I with only a small number of defects. The material is characterised by high aspect ratio, high surface area, very large modulus of elasticity (around 130 GPa) and strength in the order of 7 GPa [6]. In addition, a significant advantage of CNC is also its modifiable surface properties due to reactive –OH side groups [11]. In addition to, it has been shown that the process of biodegradability is also conditioned by the size of the material, as it is known that nano-cellulose is significantly more biodegradable compared to the same cellulosic material in macro size [11]. Recently, due to the great interest in nanocellulose, numerous articles provide detailed reviews on nanocrystalline cellulose processing, modification and application (e.g. [1, 6, 14–16]).

Nanocrystalline cellulose is obtained by dissolving away regions of low lateral order and the water-insoluble, highly crystalline residue is converted into a stable suspension by subsequent vigorous mechanical shearing action [17]. Rod-like rigid cellulose monocrystals are obtained. Nanoparticles' dimensions depend on the biological source of the substrate [17].

Typical processes currently employed for the preparation of NCC consists of subjecting cellulosic material to strong acid hydrolysis under strictly controlled conditions of temperature, time, acid concentration and acid—pulp ratio [18]. However, the process of controlled sulphuric acid—catalysed degradation of cellulose fibers for preparation of cellulose colloidal suspensions has been known since 1952, when Ränby reported about it [19]. Since then the process has been studied intensively by a number of researchers (e.g. [20, 21], etc.) and optimised (e.g. by ultrasonication) [18].

The amorphous regions absorb chemicals easily, whereas the compactness of the crystalline regions makes it difficult for chemical penetration [1, 2]. The non-crystalline region is hydrolysed preferentially, i.e. acid hydrolysis of cellulose leads to hydrolytic cleavage of the glycosidic bond between two anhydroglucose units in amorphous domains. Thus, the amorphous portion gets dissolved by acid hydrolysis, thus releasing the cellulose crystals [2]. The remaining crystals are in the nano-metre size range. Usually chemical hydrolysis with sulphuric acid is used for nanocellulose preparation, although hydrochloric acid, phosphoric acid and hydrobromic acid have been used in addition [6]. Besides, formic acid and nitric acid have been used as well. However, due to a high acid dissociation constant, HCl and H_2SO_4 possess a superior ability to hydrolyze β-1,4-glucosidic bonds [22]. The process is treatment temperature and time conditioned. Analogously to the acid hydrolysis treatment, cellulase enzymes are attacking the amorphous regions of cellulosic substrates [6].

By acid hydrolysis with sulfuric acids due to the surface charges (conversion of cellulose –OH to cellulose $-OSO_3^-H^+$), a more stable colloidal dispersion is obtained when compared to nanocellulose dispersions obtained with hydrolytical degradation by hydrochloric and hydrobromic acids [7, 23]. The introduction of charged functional groups on CNC was studied in addition by Cudjoe, and he observed that crystals with charged surfaces are more dispersible in certain polar solvents, which influence further surface functionalization [14].

Different cellulose sources were studied and the influence of different process parameters was examined i.e. treatment temperature and time, acid concentration, etc. on nano-crystal's characteristics and yield. In addition, nanocrystalline cellulose was prepared by acid hydrolysis combined with the ultrasonication method [18].

The processing parameters, i.e. the concentration of microcrystalline cellulose (MCC) and sulphuric acid, the hydrolysis time and temperature, and the ultrasonic treatment time for acid hydrolysis of MCC derived from Norway spruce [*Picea abies* (L.) Karst.], were optimised by Bondeson [21]. The average size of the cellulose nanocrystals and yield were studied. The length of nanocrystals was between 200 and 400 nm and the width was less than 10 nm after treating the samples for 2 h in sulphuric acid with a concentration of 63.5% (w/w) and the yield of the optimised process was 30% (of initial weight) [21].

For black spruce [Picea mariana Mill.], it was found that longer hydrolysis times produced shorter, less polydisperse cellulose nanocrystals and increased acid-to-pulp ratio reduced the dimensions of the nanocrystals thus produced. Very similar results were obtained for bleached kraft eucalyptus pulp when prepared under similar hydrolysis conditions [20].

Besides, for the preparation of nano-crystalline cellulose various non-wood cellulose sources, such as cotton, or cotton linters are involved. The shape and size distribution of crystalline nanoparticles prepared by acid hydrolysis of cotton and Avicel, a commercial microcrystalline cellulose, was studied by Elazzouzi-Hafraoui with co-workers. Particles were composed of a small number of laterally associated elementary crystallites (3–4 for cotton, 2–3 for Avicel) and they were polydisperse in length and width, but rather homogeneous in thickness. The length of cotton nanoparticles was between 25 and 320 nm and the width from 6 to 70 nm. The length of Avicel particles was lower than that of cotton particles and it ranged from 35 to 265 nm and the width from 3 to 48 nm [24].

5.1   Non-wood Lignocellulosic Sources for CNCs

Nanocellulose is extracted mostly from wood pulp or pure cellulose such as micro-crystalline cellulose MCC [11, 25]. Wood, because of its great abundance, is often an attractive source of cellulose for designing nanocellulose. However, multi-stage procedures are needed for isolation of cellulosic structures with nm dimensions (in the range of 1–100 nm) from wood complex structures and, therefore, more pure cellulose sources often have an advantage [6].

Therefore, a number of crop residues were proposed as the starting material for preparation of nanocrystals, such as soybean stock, swede root, sugar beet pulp, potato tubers, wheat straw [6], chardonnay grape skins [26], cassava bagasse [27], cotton stalk [28], bamboo [29, 30], agave T, barley husks [25], etc. Globally, by-products arising from cultivation of corn, wheat, rice, sorghum, barley, sugar-cane, pineapple, banana and coconut are the major sources of agro-based cellulosic fibres and, hence, CNC [31, 32]. In addition, several non-traditional cellulose fibre sources were studied as well: e.g. *Miscanthus xgiganteus* [14], switch grass, water hyacinth [33], mulberry bark [34], loofah [35], zoysia grass [36], etc.

Wheat straw was studied as a cellulose source by several authors. It was found that crystals obtained by acid hydrolysis of wheat straw are little rods of 150–300 nm in length and with a width of about 5 nm. The aspect ratio L/d is 50 [29, 37]. Furthermore, wheat straw was researched for cellulose nanofibres' isolation by

Alemdar [30]. Chemi-mechanical processes were applied for preparation of parti-cles with diameters in the range of 10–80 nm and lengths of several thousand nanometres [30].

Cellulose nanofibres from water hyacinth cellulose were prepared successfully by Marimuthu [33]. Water hyacinth [*Eichhornia crassipes* (Mart.) Solms] is a fast growing invasive perennial aquatic weed which reproduces primary by stolons. The noxious weeds double their population in 2 weeks and are rich in fibre content. Bleaching and alkaline and sodium chlorite reactions were involved for preparation of nano-cellulose. A diameter in the range between 20 and 100 nm from the SEM and 25 nm from the TEM analysis was defined for the obtained nanoparticles. Subsequently, extracted cellulose nanofibres were applied for preparation of biodegradable nanocomposites in drug delivery, ligament substitute, filtration, and water purification and also used as flexible optical displays in electronics [33].

Nanocrystallites from ramie cellulose were generated by acid hydrolysis for designing composites. Cellulose nanocrystalites were prepared with average lengths of 538.5 ± 125.3 nm and diameters of 85.4 ± 25.3 nm. The results indicate that the synergistic interactions between fillers and between filler and matrix play a key role in reinforcing the composites [38].

Different traditional cellulose textile fibres were also examined for nanocellulose preparation by acid-catalyzed hydrolysis. Cao prepared cellulose nanocrystals by acid hydrolysis of flax fibre [39]. Flax nanocrystals are slender rods with a broad distribution in size. Particles' average length of 327 ± 108 nm and diameter of 21 ± 7 nm, were determined respectively [39]. Particles were used as reinforce-ment for nanocomposite films with polycaprolactone-based waterborne poly-urethane as the matrix [39] and to reinforce the plasticized starch (PS) matrix [40].

One of the cellulose sources studied for the preparation of nano-celluloses is also mandarin residue that is left by citrus juice processing. The production of citrus juice on an industrial level leads to a considerable quantity of solid and liquid residues (globally around 8–20 million tons per year), which is considered waste. Cellulose nanofibrils with thicknesses of 2–3 nm were successfully obtained from satsuma mandarin [*Cistrus unshiu* (Swingle) Marcow] [41].

First trials were made to obtain cellulose nanocrystals from chardonnay grape skins. The process of cellulose isolation included several steps, i.e. organic extraction, acid and base dissolutions, and basic and acidic oxidation. Highly crystalline cellulose (54.9%) with the cellulose yield of 16.4% was obtained. Cellulose nanocrystals with the crystallinity 64.3% were mostly spherical. The mean diameter of particles was 48.1 ± 14.6 nm [26].

By H_2SO_4 hydrolysis of purified soy hulls at 40 °C for 30 min and 40 min, respectively, aggregates of nanocrystals and individual needle-like particles were obtained and analysed by Neto et al. [3]. The crystallinity of nanocellulose, obtained by the procedure when acid hydrolysis was performed for 30 min, was 73.5%. The average cross-sectional dimension of the elementary cellulose crys-tallites for nanocrystals was 2.73 nm. However, a decrease in crystallinity was observed when the extraction time was longer. The authors argue that, in this case, not only the amorphous portion of cellulose is removed, but also some crystalline domains are partly destroyed [3]. In addition, some changes in the crystalline structure of treated samples were observed by the same researchers. Namely, for untreated samples, a predominance of cellulose I crystalline structure was confirmed by the presence of peaks at $2\theta = 15°$ (plane 101), $17°$ (plan 10-1), $21°$ (plane 021), $23°$ (plane 002) and $34°$ (plane 004). However, by the hydrolytical treatment, a mixture was obtained of polymorphs of cellulose I and cellulose II. The presence of cellulose II was confirmed by the shift of the maxima to $2\theta = 12°$ (plane 101), $20°$ (plan 10-1) and $22°$ (plane 002) [3].

Teixeira, with co-workers, extracted cassava cellulose bagasse nanofibres (CBN) from cassava bagasse (CB) cellulose fibres. The diameter and length of the CBN is in the range of 2–11 nm and 360–1700 nm, respectively. XRD analysis for CB fibres and CB nanofibres shows two well-defined peaks typical for cellulose I ($2\theta = 12.5°$ and $2\theta = 22.5°$). Both peaks are narrower, and increased by acid treatment of fibres due to increased crystallinity of the samples [42]. The measured crystallinity index was 43.7 and 54.1% for the CB and CBN, respectively. The authors suppose that the presence of other compounds than nanocellulose and the remains of amorphous cellulosic domains are the reason for the relatively low degree of crystallinity of the hydrolyzed CB [27, 42].

Cellulose whiskers were also prepared from branch-barks of white mulberry (*Morus alba*) by an alkali treatment at 130 °C and, subsequently, a sulphuric acid hydrolysis. The diameter of the extracted whiskers ranged from 20 to 40 nm and the crystallinity was 73.4% [34].

Miscanthus x giganteus pulp obtained by alkali treatment and bleaching was used for extraction of cellulose nanocrystals. Therefore, hydrochloric acid and sulphuric acid hydrolysis was used. By HCl hydrolysis, high crystalline samples were formed with crystallinity index 94% [14]. Studies of the preparation of CNC from MG has demonstrated a high overall yield of 33% (w/w) of the CNC with relatively high aspect ratios (60–70) from MG stalks after base treatments, bleaching and acid hydrolysis. The crystals were ribbon-like with a width and thickness of 8.5 and 2.8 nm, respectively [14].

Dried sugar beet pulp chips are a by-product of the sugar industry. Sugar beet pulp primary cell walls consist of approximately one third cellulose, one third hemicelluloses, and one third pectin [31, 43]. Leitner, with co-authors, isolated cellulose fibrils from cells, which are nano-scaled, with diameters ranging from 30 to 100 nm and with a length of several mm. Nanofibrils were studied by different techniques (e.g. Wide-Angle X-ray Scattering WAXS) and used for preparation of nanocomposites with PVA.

Crystalline nanocellulose based on cellulose type I was produced from blue agave (*Agave tequilana*), and compared to particles obtained from micro-crystalline cellulose MCC and barley husks [25]. Length of rod like particles was 218 ± 60 nm, 323 ± 113 nm and 329 ± 123 nm, respectively for CNC extracted from MCC, agave T and barley respectively. Particles' width was 16 ± 5 nm, 11 ± 4 nm and 10 ± 4 nm as well as aspect ratio 13, 28 and 32 for MCC, agave T and barley husks CNC. Particles were analysed by dynamic light scattering for evaluation of particles sizes. Nanocrystals' sizes were: 188 ± 13 nm,

214 ± 19 nm and 269 ± 16 nm for CNC from MCC, agave T and barley husks. CNC from agricultural by-products (agave T x_{cr} = 71% and barley husk x_c = 66) show a lower crystallinity index than CNC from MCC (87%) [25].

Cellulose nanocrystals were also extracted from *Agave augustifolia* (Haw.) fibres by alkali and bleaching treatments, followed by acid hydrolysis. The diameter of the nano-crystalline cellulose particles ranged from 8 to 15 nm and 170 to 500 nm in length. The average diameter of CNC was 10 nm, the length 310 nm and the crystallinity of particles 82% [44].

Barley husks were studied for their utilisation for CNC preparation by Espino et al. [25]. They produced CNC successfully with length and diameter 329 ± 123 nm and 10 ± 4 nm, respectively. Nano crystals were highly crystalline (66%) and appropriate for nanocomposite preparation [25].

Loofah (*Luffa cylindrica*) was used as a source for nanocellulose particles' isolation by the acid hydrolysis method. CNC was highly crystalline (96.5%). Particles' dimensions were: length 242 ± 86 nm and diameter 5.2 ± 1.3 nm [35].

Chen and co-workers have prepared pea hull fibre (PHF)-derived nanocellulose successfully by sulphuric acid hydrolysis of PHF for their utilisation as a filler in nanocomposite films with pea starch [45]. The group studied the effect of nanocrystals hydrolyzed from PHF under different conditions on the properties of the pea starch-based composite films when they were used as a filler in a pea starch-matrix. It was found that crystals hydrolyzed from PHF by sulphuric acid were needle-like or rod-like. The average dimensions of nanocellulose showed a dependence on the hydrolysis time. The lengths of crystallites were from 400 to 240 nm, and diameters from 12 to 7 nm dependent on hydrolysis time (4–24 h). An XRD diagram of PHF exhibited the typical cellulose I pattern with peaks at about 2θ = 22.2°, 30.7° and 34.7°, and a broad peak around 15.3°, arising from the amorphous structures. By acid treatment, the peak at 2θ = 22.5° is stronger and narrow, and there are additional relatively lower peaks at 2θ = 14.9°, 16.3° and 34.6°. The sharper diffraction peak is an indication of a higher degree of crystallinity in the nano-cellulose structure [45].

Cellulose nanocrystals (CNC) were successfully extracted by hydrolysis from okra (*Abelmoschus esculentus*) bast fibers for preparing PVA based nanocomposites [46].

A group of Korean researchers isolated nanocrystals from zoysia [36]. Samples were hydrolyzed for 15 min at 55 °C with 65% H_2SO_4 solution and dyalized. The dialyzed product was freeze-dried, followed by sonication. Needle-shaped nano-entities were obtained with a length between 180 and 300 nm and width less than 10 nm. The crystalline structure of isolated grass nano-crystals was studied by X-Ray Diffraction XRD. Some changes were observed on the X-ray diffraction diagram of nano-crystalline cellulose compared to the diagram of isolated fibres. Two additional peaks were observed at 2θ = 15.5° and 2θ = 16.2° for the (110) and (-1-10) plane. From the sharper nanocellulose peaks the authors concluded that nanocellulose had higher crystallinity when compared to the crystallinity of isolated fibres [36]. Based on cellulose nanocrystals, obtained from zoysia grass, the composites were fabricated with poly(lactic acid) in the presence of a compatibilizer [47].

It has often been proven that the structure and properties, especially the size of the nanocellulose, depend on the source of the original cellulose materials, and on the preparation process. Properties of nanocellulose, dependent on the starting cellulose substrate, are collected in Table 5.1.

Table 5.1 Properties of nanocellulose from different plant origins

Fiber type	Crystallites diameter (nm)	Crystallites length (nm)	L/d	References
Micro—crystalline cellulose MCC	16 ± 5	218 ± 60	13	Espino [25]
Cotton	8–10	100–300		de Souza Lima [12]
Sisal	2–11	360–1700		Martelli-Tosi [48]
Wheat straw	30–40	>100		Martelli-Tosi [48]
	5	150–300		Dufresne [29]
Hemp	20–40	100–500		Martelli-Tosi [48]
Flax	10–30	100–500		Cao [40]
Cassava bagasse	2–11	360–1700		Teixeira [42]
Pea hull	7–12	240–400	34	Chen [45], Martelli-Tosi [48]
Agave tequilana	11 ± 4	323 ± 113	28	Espino [25]
Barley husks	10 ± 4	329 ± 123	32	Espino [25]
Soy hull	4.43 ± 1.20	122.7 ± 39.40	29.41 ± 11.53	Neto [3]
MG	2.8–8.5			Cudjoe [14]
Mulberry	20–40			Li [34]
Zoysia grass	<10	180–300		Pandey [36]
Water hyacinth	20–100			Marimuthu [33]
Ramie	85.4 ± 25.3	538 ± 125.3		Lu [38]
Mandarin residue	2–3			Hiasa [41]
Agave augustifolia	10	310	10–45	Rosli [44]
Luffa cylindrica	5.2 ± 1.3	242 ± 86	46.8	Siqueira [35]
Mengkuang leaves (Pandanus tectorius)	5–25	200	10–20	Sheltami [13]
Grape skin	48.1 ± 14.6 spherical			Lu [26]

References

1. Klemm D, Kramer F, Moritz S et al (2011) Nanocelluloses: a new family of nature-based materials. Angew Chem Int Ed 50:5438–5466
2. Abraham E, Deepa B, Pothan LA et al (2011) Extraction of nanocellulose fibrils from lignocellulosic fibres: a novel approach. Carbohyd Polym 86:1468–1475
3. Neto WPF, Silvério HA, Dantas NO et al (2013) Extraction and characterization of cellulose nanocrystals from agro-industrial residue—Soy hulls. Ind Crops Prod 42:480–488
4. Orts WJ, Shey J, Imam SH et al (2005) Application of cellulose microfibrils in polymer nanocomposites. J Polym Environ 13:301. https://doi.org/10.1007/s10924-005-5514-3
5. Khalil HPSA, Bhat AH, Abu Bakar A et al (2014a) Cellulosic nanocomposites from natural fibers for medical applications: a review In: Padney JK (ed) Handbook of polymer nanocomposites. Processing, performance and application, Volume C: Polymer nanocomposites of cellulose nanoparticles, Springer Verlag, Berlin, Heidelberg, p 475–511. https://doi.org/10.1007/978-3-642-45232-1_72
6. Hubbe MA, Rojas OJ, Lucia LA et al (2008) Cellulosic nanocomposites: a review. Bioresources 3:929–980
7. Börjesson M, Westman G (2015) Crystalline nanocellulose—preparation, modification, and properties, cellulose—fundamental aspects and current trends. In: Poletto M (ed) InTech, https://doi.org/10.5772/61899
8. Oksman K (2012) Nanocelluloses and their use in composite materials. Express Polym Lett 6 (9):687
9. Nechyporchuk O, Belgacem MN, Bras J (2016) Production of cellulose nanofibrils: a review of recent advances. Ind Crops Prod 93:2–25
10. Khalil HPSA, Davoudpour Y, Nazrul Islam MD (2014) Production and Modification of nanofibrillated cellulose using various mechanical processes: a review. Carbohyd Polym 99:649–665
11. Brinchi L, Cotana F, Fortunati E et al (2013) Production of nanocrystalline cellulose from lignocellulosic biomass: technology and Applications. Carbohyd Polym 94:154–169
12. de Souza Lima MM, Borsali R (2004) Rodlike cellulose microcrystals: structure, properties, and applications. Macromol Rapid Commun 25:771–787. https://doi.org/10.1002/marc.200300268
13. Sheltami RM, Abdullah I, Ahmad I et al (2012) Extraction of cellulose nanocrystals from mengkuang leaves (Pandanus tectorius). Carbohyd Polym 88(2):772–779
14. Cudjoe E, Hunsen M, Xue Z et al (2017) *Miscanthus Giganteus*: a commercially viable sustainable source of cellulose nanocrystals. Carbohyd Polym 155:230–241. https://doi.org/10.1016/j.carbpol.2016.08.049
15. Eichhorn SJ, Dufresne A, Aranguren M et al (2010) Review: current international research into cellulose nanofibres and nanocomposites. J Mater Sci 45:1–33. https://doi.org/10.1007/s10853-009-3874-0
16. Moran JI, Alvarez VA, Cyras VP et al (2008) Extraction of cellulose and preparation of nanocellulose from sisal fiber. Cellulose 15(1):149–159
17. Dufresne A (2006) Comparing the mechanical properties of high performances polymer nanocomposites from biological sources. J Nanosci Nanotechnol 6(2):322–630
18. Li W, Wang R, Liu S (2011) Nanocrystalline cellulose prepared from softwood kraft pulp via ultrasonic—assisted acid hydrolysis. BioResources 6(4):4271–4281
19. Habibi Y, Lucia LA, Rojas OJ (2010) Cellulose nanocrystals: chemistry, self-assembly and applications. Chem Rev 110(6):3479–3500
20. Beck-Candanedo S, Roman M, Gray DG (2005) Effect of reaction conditions on the properties and behavior of wood cellulose nanocrystal suspensions. Biomacromol 6(2):1048–1054

21. Bondeson D, Mathew A, Oksman K (2006) Optimization of the isolation of nanocrystals from microcrystalline cellulose by acid hydrolysis. Cellulose 13:171. https://doi.org/10.1007/s10570-006-9061-4
22. Sri Aprilia NA, Davoudpour Y, Zulqarnain W et al (2016) Physicochemical characterization of microcrystalline cellulose extracted from kenaf bast. BioResources 11(2):3875–3889
23. Peng BL, Dhar N, Liu HL et al (2011) Chemistry and applications of nanocrystalline cellulose and its derivatives: a nanotechnology perspective. Can J Chem Eng 9999:1–16
24. Elazzouzi-Hafraoui S, Nishiyama Y, Putaux J et al (2008) The shape and size distribution of crystalline nanoparticles prepared by acid hydrolysis of native cellulose. Biomacromolecules 9(1):57–65
25. Espino E, Cakir M, Domenek S et al (2014) Isolation and characterization of cellulose nanocrystals from industrial by-products of Agave tequilana and barley. Ind Crops Prod 62:552–559
26. Lu P, Hsieh YL (2012) Cellulose isolation and core-shell nanostructures of cellulose nanocrystals from chardonnay grape skins. Carbohyd Polym 87(4):2546–2553. https://doi.org/10.1016/j.carbpol.2011.11.023
27. Pasquini D, Teixeira EM, da Silva Curvelo AA (2010) Extraction of cellulose whiskers from cassava bagasse and their applications as reinforcing agent in natural rubber. Ind Crops Prod 32:486–490
28. Habibi Y, El-Zawawy WK, Ibrahim MM et al (2008) Processing and characterization of reinforced polyethylene composites made with lignocellulosic fibers from Egyptian agroindustrial residues. Compos Sci Technol 68(7–8):1877–1885. https://doi.org/10.1016/j.compscitech.2008.01.008
29. Dufresne A, Cavaille JY, Helbert W (1997) Thermoplastic nanocomposites filled with wheat straw cellulose whiskers, effect of processing and modeling. Polym Compos 18(2):198–210
30. Alemdar A, Sain M (2008) Biocomposites from wheat straw nanofibers: morphology, thermal and mechanical properties. Compos Sci Technol 68(2):557–565
31. Leitner J, Hinterstoisser B, Wastyn M et al (2007) Sugar beet cellulose nanofibril-reinforced composites. Cellulose 14:419–425
32. Reddy N, Yang Y (2005) Biofibers from agricultural byproducts for industrial applications. Trends Biotechnol 23(1):22–27
33. Marimuthu TS, Atmakuru R (2015) Isolation and characterization of cellulose nanofibers from the aquatic weed water hyacinth: eichhornia crassipes. In: Pandey JK, Takagi H, Nakagaito AN et al (eds) Handbook of polymer nanocomposites. Processing, performance and application; Volume C: Polymer nanocomposites of cellulose nanoparticles. Springer Verlag, Berlin, Heidelberg, pp 37–46
34. Li R, Fei J, Cai Y et al (2009) Cellulose whiskers extracted from mulberry: a novel biomass production. Carbohyd Polym 76(1–2):94–99
35. Siqueira G, Bras J, Dufresne A (2010) Luffa cylindrica as a lignocellulosic source of fibre, microfibrillate cellulose and cellulose nanocrystals. BioResources 5(2):727–740
36. Pandey JK, Lee JW, Chu WS et al (2008) Cellulose nano whiskers from grass of Korea. Macromol Res 16(5):396–398
37. Helbert W, Cavaille JY, Dufresne A (1996) Thermoplastic nanocomposites filled with wheat straw cellulose whiskers. Part I: processing and mechanical behavior. Polym Compos 17(4):604–611
38. Lu Y, Weng L, Cao X (2006) Morphological, thermal and mechanical properties of ramie crystallites—reinforced plasticized starch biocomposites. Carbohyd Polym 63(2):198–204
39. Cao X, Dong H, Li CM (2007) New nanocomposite materials reinforced with flax cellulose nanocrystals in waterborne polyurethane. Biomacromol 8(3):899–904
40. Cao X, Chen Y, Chang PR et al (2008) Starch-based nanocomposites reinforced with flax cellulose nanocrystals. Express Polymer Letters 2(7):502–510
41. Hiasa S, Iwamoto S, Endo T et al (2014) Isolation of cellulose nanofibrils from mandarin (Citrus unshiu) peel waste. Ind Crops Prod 62:280–285

42. Teixeira EM, Pasquini D, Curvelo AAS (2009) Cassava bagasse cellulose nanofibrils reinforced thermoplastic cassava starch. Carbohyd Polym 78:422–431

43. Dinand E, Chanzy H, Vignon RM (1999) Suspensions of cellulose microfibrils from sugar beet pulp. Food Hydrocolloids 13(3):275–283

44. Rosli N, Ahmad I, Abdullah I (2013) Isolation and characterization of cellulose nanocrystals from Agave angustifolia fibre. BioResources 8(2):1893–1908

45. Chen Y, Liu C, Chang PR (2009) Bionanocomposites based on pea starch and cellulose nanowhiskers hydrolyzed from pea hull fibre: effect of hydrolysis time. Carbohyd Polym 76:607–615

46. Fortunati E, Puglia D, Monti M et al (2012) Cellulose nanocrystals extracted from okra fibers in PVA nanocomposites. J Appl Polym Sci 128(5):3220–3230. https://doi.org/10.1002/app.38524

47. Pandey JK, Chu WS, Kim CS et al (2009) Bio-nano reinforcement of environmentally degradable polymer matrix by cellulose whiskers from grass. Compos B 40(7):676–680

48. Martelli-Tosi M, da Silva Torricillas M, Martins MA et al (2016) Using commercial enzymes to produce cellulose nanofibers from soybean straw. J Nanomaterials 2016. https://doi.org/10.1155/2016/8106814

Chapter 6
Preparation of Cellulose Nanocrystals CNC from Nettle, Weeping Willow, Balm-Leaved Archangel, Lucerne and Spanish Broom

Abstract Some plants growing wild in continental and mediterranean European climates were studied as potential sources of cellulose fibres, i.e. nettle, weeping willow, balm-leaved archangel, lucerne and Spanish broom. Extraction of fibres by enzyme retting and chemical retting, respectively, fibres' isolation and pre-treatment are described. In addition to fibres properties were determined and finally CNC was prepared from isolated cellulose fibres. Nano-crystalline cellulose was characterised by SEM observations and particles dimensions were determined.

Keywords Cellulose nanocrystals CNC · Nettle · Weeping willow · Balm-leaved archangel · Lucerne · Spanish broom · Morphology

Some plants growing wild in continental and mediterranean European climates were studied by our group as potential sources of cellulose fibres. Plants for observation were collected in Slovenia and Croatia in 2016. Nettle (*Urtica dioica L.*), weeping willow (*Salix babylonica L.*), balm-leaved archangel (*Lamium orvala L.*), lucerne (*Medicago sativa L.*) and hemp (*Cannabis sativa* L.) were collected in the vicinity of Ljubljana with the geographical latitude and longitude 46°03′20″N and 14°30′30″E, respectively. Plants were grown at an altitude of about 300 m. More exactly, weeping willow, lucerne, balm-leaved archangel and common nettle have been harvested in April and May 2016 in the area of Kostanjevica (45°50′55.2″N; 15°25′02.3″E) and hemp fibres were obtained from Begunje, Gorenjska region (46°22′35.1″N;14°12′22.3″E). Spanish broom was obtained from the mediterranean region (Zadar; 44°07′10″N; 15°14′31″E).

Fibres were extracted from plant stems. They lie along the length of the stem beneath the surface of the outer bast layer (Fig. 6.1). Different procedures for fibres' isolation from stems were involved and morphology of fibres and their mechanical properties were studied. In addition, isolated fibres were used for preparation of cellulose nanocrystals. Fibres' properties were compared to hemp fibres grown in the same region as other fibres.

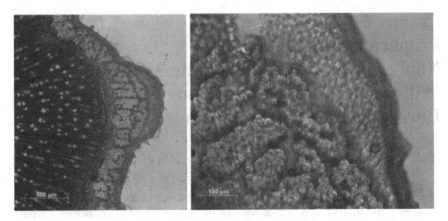

Fig. 6.1 Cross-section of a hemp stem with fibres' bundles

6.1 Extraction of Fibres

Depending on how strong the connection between fibrous cells and the cells of the surrounding tissues were, we performed different processes of retting, i.e. the process of chemical and biochemical retting, respectively, were used for fibres' extraction. Chemical reagents such as NaOH and H_2O_2, which were used for the chemical pretreatment of stems, were of analytical grade.

Enzyme retting

Retting was performed in a Labomat (Werner Mathis AG, Switzerland) dyeing apparatus according to the procedure proposed by the enzyme producer Bezema. The non-ionic surfactant Sandoclean (2 w/w%) (Clariant Switzerland) and enzymes Beisol PRO (4 w/w%) were used for retting solution preparation. Bezema (Switzerland) provided Beisol PRO enzymes. The liquor ratio of the treatment solution was 1:15 and pH 8–9, that was adjusted by NaOH (Sigma-Aldrich, USA).

Leaves and flowers were removed from the stems, which were then cut into smaller pieces (10–15 cm) and weighed. The volume of the bath was prepared according to the liquor ratio. Stems were placed into the Labomat dye beakers filled with the enzyme solution. The treatment conditions were set on: T = 55 °C, t = 15 min and 4.0 rpm. After 15 min, the temperature was increased to 85 °C. The treatment was performed at this condition for additional 90 min. The samples were washed with distilled water (T = 60 °C) and then with tap water. The samples were dried on a filter paper, as shown in Fig. 6.2. Enzymatic retting was used for nettle, Spanish broom, balm-leaved archangel and lucerne.

Fig. 6.2 Retted stems of nettle, Spanish broom, balm-leaved archangel and lucerne

Chemical retting

Retting was performed in an alkaline NaOH solution (15 w/w%; Sigma-Aldrich, USA) in a Labomat (Werner Mathis AG, Switzerland) dyeing apparatus. Stems were cut into pieces (10–15 cm) and purified and inserted into Labomat dye beakers. The liquor ratio was 1:10. The necessary volume of the treatment solution was added conditioned by stems' weight. Dyeing apparatus was set on the following working parameters: T = 120 °C, t = 3 h, 4.0 rpm. After the chemical treatment was finished, stems were washed in a strainer with distilled water and dried on a filter paper in air. Tree branches of weeping willow were retted according to this chemical procedure.

Fibres' isolation

Fibre separation from non-cellulose cells was performed by hand, with preparation needles. The retted stems were placed on a dark basis, and fibres were separated manually from each other by needles. If necessary, stems were wetted slightly. For further analyses, fibres were conditioned for 24 h under standard climate conditions (T = 20 °C ± 2 °C; RH = 65% ± 2%) in a climate chamber.

Fibres' pre-treatment

For preparation of CNC it was necessary to remove all non-cellulosic components from the fibres and, therefore, we bleached them. The peroxyde bleaching process was performed in a Labomat dyeing apparatus at the liquor ratio 1:50. Bleaching solution was prepared with the addition of Tannex GEO (2 mL/L). Tannex GEO (Tanatex, Netherlands) is a nonionic stabilizer based on natural minerals for discontinuous hydrogen peroxide bleaching. The process was performed at pH = 10.7 with 6 mL/L H_2O_2 (Carlo Erba Reagents, Italy). pH was adjusted by NaOH (Sigma-Aldrich, USA).

Fibre samples were placed into Labomat cuvettes. Treatment was performed at the temperature of 98 °C, for 30 min and at 4.0 rpm. After 30 min we took samples from the Labomat and washed them. Fibres were placed into beakers filled with distilled water, and placed on magnetic stirrers. After 5 min the distilled water was changed and, after an additional 5 min, the electrical conductivity of the water was measured. Fibres were finally washed when electrical conductivity was lower than 30 µS. Finally, fibres were dried for 24 h at 80 °C.

Additional extraction of samples was performed in ethanol (96 w/w%) in a Soxhlet extractor. First, samples were wrapped in filter paper and placed in the extractor and extracted for 6 h in 200 mL ethanol (Sigma-Aldrich, USA). Subsequently, samples were washed and dried on a filter paper.

6.2 Fibres' Properties

6.2.1 Mechanical Properties

Extracted and pre-treated fibres were analysed. Fibres' mechanical properties were measured according to the Standard method ISO 5079:1995 (ISO 5079:1995 Textile fibres—Determination of breaking force and elongation at break of individual fibres) and linear density was evaluated according to the Standard method ISO 1973: 1995 (ISO 1973: 1995 Textile fibres—Determination of linear density—Gravimetric method and vibroscope method). Vibrodyn and Vibroskop (Lenzing, Austria) were used for measuring fibres' tenacity and fineness. Measurements were performed under standard conditions (T = 20 °C and relative RH = 65%), 10 parallel samples were measured. Results are given in Table 6.1.

High deviations of the results of fineness and tenacity measurements are observed. Such variations are typical for the multi-cellular vegetable fibers. Fibres fineness is conditioned by the arrangement and number of fiber cells in bundles and by the fineness of ultimate cells, but also by their isolation procedure. The finest

Table 6.1 Linear density and mechanical properties of nettle (*Urtica dioica*), Spanish broom (*Spartium junceum*), balm-leaved archangel (*Lamium orvala*) and lucerne (*Medicago sativa*), weeping willow (*Salix babylonica*) and hemp (*Cannabis sativa*)

Fibre type	Fineness (dtex)	Variation coefficient VC fineness (%)	Tenacity cN/tex	Variation coefficient VC tenacity (%)	Elongation (%)	Variation coefficient VC elongation (%)
Nettle	32.6	24.2	58.1	20.8	3.6	26.7
Spanish broom	29.9	19.8	41.3	21.7	4.7	21.4
Balm-leaved archangel	32.4	24.9	19.8	27.5	4.2	17,8
Lucerne	58.7	22.8	37.3	83.2	4.8	22
Weeping willow	27.2	27.9	30.4	33.8	3.9	30.2
Hemp	19.3	36.7	36.4	32.6	3.2	34.6

fibers were isolated from the stems of hemp (19.3 dtex) and the coarsest from lucerne (58.7 dtex). Nettle fibers are characterized by the highest breaking strength properties, whereas the fibers obtained from balm-leaved archangel have a minimum tenacity. The results of the measurements were compared to values obtained by some other researchers. Östbom with co-workers have analysed several hemp fibres obtained by water and dew retting. Fibres' tenacities were in the range from 38.4 to 48.5 cN/tex and are conditioned by species, retting and growth conditions; e.g. climate, soil [1]. Sankari researched fibre yield and mechanical properties of hemp cultivars. The median values for breaking tenacity and elongation at break of hemp fibers varied, depending on experimental year and cultivar, from 41 to 74 cN/ tex and from 3.3 to 5.5%, respectively [2]. Mean tenacity 24.20 cN/tex of unmodified single hemp fibers (200 ultimate fibres were tested and the coefficient of variation was 30.1%) was determined by Kostic et al. [3].

For nettle fibres the moisture content of 12.5%, fineness in the range of 1.0 tex and 1.2 tex as well as dry tenacity of 5–7 cN/tex at a breaking extension of 2.1– 2.3% was obtained by Tadele et al. [4]. Mechanical properties were significantly influenced by the fibres age [4]. Properties of fibres obtained by different isolation methods from Spanish broom were studied by several researchers, e.g. Cerchiara, Kovacevic etc. Fibres' tenacity was determined in the range between 10 and 70 cN/ tex; i.e. 64.44 cN/tex [5] and 35.9 cN/tex [6], respectively depending on the isolation method and pre-treatment procedures. The results of our group were in good agreement with the results of other researchers and reflect that due to the properties of all analyzed fibres they are suitable for use for different applications.

6.2.2 Morphological Properties

The morphology of extracted and pre-treated fibres was observed on a Scanning Electron Microscope SEM Carl Zeiss FE-SEM Supra 35 VP at U = 1 kV. Figures 6.3, 6.4, 6.5, 6.6, 6.7 and 6.8 represent SEM images of the observed fibres.

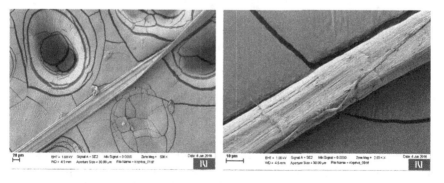

Fig. 6.3 SEM images (500 and 2500×) of nettle (*Urtica dioica*)

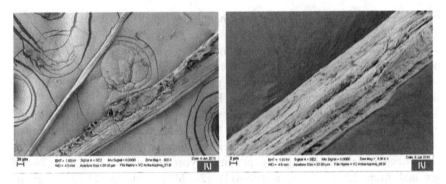

Fig. 6.4 SEM images (500 and 5000×) of balm-leaved archangel (*Lamium orvala*)

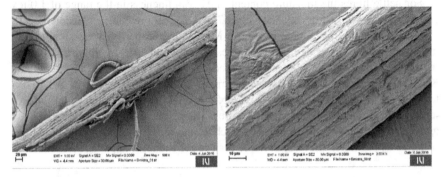

Fig. 6.5 SEM images (500 and 2500×) of Spanish broom (*Spartium junceum)*

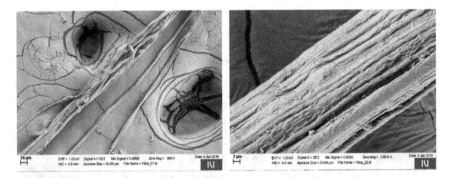

Fig. 6.6 SEM images (500 and 5000×) of weeping willow (*Salix babylonica*)

The typical morphological characteristics of stem fibers are observed for all different types of extracted stem fibres. Fibers are bundles of ultimate cells, flat and untwisted. Lateral lines are reflecting the boundaries between the individual fibre cells. Thickness of fibers isolated from Spanish broom and lucerne is about 40 and

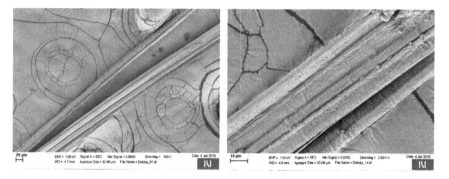

Fig. 6.7 SEM image (500 and 2500×)) of lucerne (*Medicago sativa*)

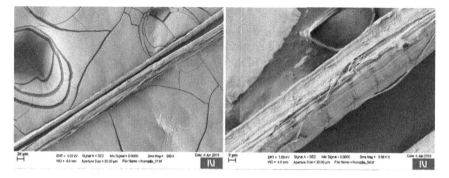

Fig. 6.8 SEM image (500 and 2500×)) of hemp (*Cannabis sativa*)

30 μm of fibers extracted from hemp and balm-leaved archangel stems. The finest fibres were isolated from willows and nettle with the thickness of 25 μm. Thickness of fibers isolated from stems is sensitive to the isolation procedure. Östbom isolated fibres from Swedish hemp stems by water and dew retting procedure and fibres' diameter was 26.6–28.7 μm and 22.2–25.9 μm, respectively [1].

6.2.3 Structural Characteristics

X-ray analysis was performed to study the crystalline structure of fibres. The scattering intensities were measured on an X-ray diffractometer Siemens D5000, operating with $Cu_{K\alpha}$ radiation ($\lambda = 15.406$ nm) measuring with a 0.04° step and a 5 s interval. The measurements were carried out from $2\theta = 10°$ to $2\theta = 40°$.

A typical scattering pattern was observed for natural cellulose I crystalline structure. A broad equatorial diffraction maximum at the 2θ angle of about 15.5° and a stronger and narrower diffraction maximum at an angle of 22° are present on the diffraction diagram. The peak at $2\theta = 22°$ is the characteristic position of a distinct and independent reflection 002 crystalline plane, however 101 and 10-1 reflections are overlapped. All analysed plant fibres show a similar XRD pattern, which characterises the crystalline structure of cellulose I allomorph. XRD diagrams are presented in Fig. 6.9.

Several techniques are suggested for evaluating the crystallinity index in the cellulosic materials from an X-ray diffractogram. For analysed fibres the crystallinity index (x_{cr}) was determined from the diffraction diagram according to the X-ray diffraction peak height method [7]. It is an empirical method and, because it is a simple method to determine the degree of crystallinity, it is used commonly to study the crystallinity of native cellulose. In this approach, the X-ray apparent crystallinity (%) of cellulose is calculated from the height ratio between the intensity of the crystalline peak and the total intensity after the subtraction of the background signal (non-crystalline), measured without cellulose according to the following equation (Eq. 6.2) [7–9]:

$$
\begin{aligned}
&X_{cr} = 100 \cdot (I_{002} - I_{AM})/I_{002}\,(\%) \\
&X_{cr} = \text{crystallinity index} \\
&I_{002} = \text{intensity of the 002 maximum} \\
&I_{AM} = \text{intensity of the amorphous scattering}
\end{aligned}
\tag{6.2}
$$

The intensity of the 002 interference was measured at $2\theta = 22.5°$ and the amorphous scatter at $2\theta = 18°$. Apparent crystallinity (%) x_{cr} was calculated from the height ratio between the intensity of the crystalline peak ($I_{002} - I_{AM}$) and total intensity (I_{002}). This method is useful for comparing the relative differences between samples. Evaluated crystallinity indices are collected in Table 6.2.

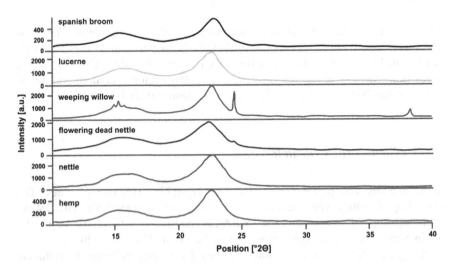

Fig. 6.9 X ray diffraction diagrams of nettle (*Urtica dioica*), Spanish broom (*Spartium junceum*), balm-leaved archangel (*Lamium orvala*) and lucerne (*Medicago sativa*), weeping willow (*Salix babylonica*) and hemp (*Cannabis sativa*)

Table 6.2 Apparent crystallinity (%) x_{cr} of alkaline treated and bleached fibres extracted from nettle (*Urtica dioica*), Spanish broom (*Spartium junceum*), balm-leaved archangel (*Lamium orvala*) and lucerne (*Medicago sativa*), weeping willow (*Salix babylonica*) and hemp (*Cannabis sativa*)

Fibre type	x_{cr} (%)
nettle (*Urtica dioica*)	77
Spanish broom (*Spartium junceum*)	70.8
balm-leaved archangel (*Lamium orvala*)	64
lucerne (*Medicago sativa*)	70.2
weeping willow (*Salix babylonica*)	80
hemp (*Cannabis sativa*)	79.2

6.3 Preparation of CNC

Nanocrystalline cellulose was prepared by strong acid hydrolysis of fibres from different plant origins after all non-cellulosic impurities had ben removed completely. The process was performed under strictly controlled conditions of temperature, time, stirring, and acid to cellulose ratio. Hydrolytic degradation in 65% H_2SO_4 solution was carried out (Carlo Erba Reagents, Italy). The liquor ratio was 1:50. The mixture was then held at 55 °C for 30 min in a ThermoHaake SWB25 thermostat (Sigma-Aldrich, USA). The samples were then centrifuged for 30 min at 8000 rpm in a Rotina 380R (Hettich, Germany) centrifuge. To stop the reaction, the suspension was diluted with an equal volume of water and washed by successive centrifugations until a turbid supernatant was observed. The nanocrystalline suspension was then purified through dialysis against distilled water to remove free acid molecules fully until neutrality. Finally, freeze-drying of the suspensions was carried out to yield solid CNC.

6.3.1 Surface Morphology of Nano-Crystalline Cellulose

For SEM observations of CNC a drop of particles' dispersion was dried at T = 50 °C on a golden tape and then after the water had evaporated the particles were observed on the scanning electron microscope Carl Zeiss FE-SEM Supra 35 VP operating at 1 kV. Figures 6.10, 6.11, 6.12, 6.13, 6.14 and 6.15 show the surface morphologies of nano-particles obtained from different plants. Nano-crystals from different origins have similar morphology. They all have an elongated rod-like structure with reduced thickness toward the ends.

Nanocrystals extracted from lucerne, nettle, weeping willow and of Spanish broom have a thickness of about 20–30 nm, and those obtained from hemp and balm-leaved archangel a thickness of about 10–20 nm. The length of nano-crystals is 100–600 nm.

Fig. 6.10 SEM image of CNC extracted from nettle (*Urtica dioica*), (the magnification 10.000 and 60.000, respectively)

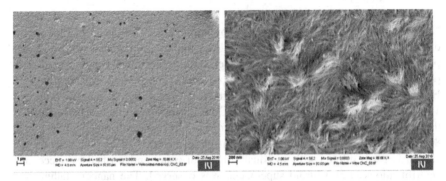

Fig. 6.11 SEM image of CNC extracted from balm-leaved archangel (*Lamium orvala*), (the magnification 10.000 and 60.000, respectively)

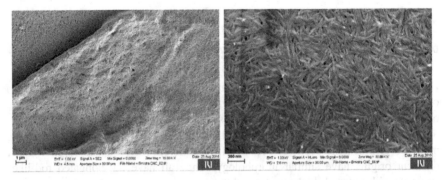

Fig. 6.12 SEM image of CNC extracted from Spanish broom (*Spartium junceum*), (the magnification 15.000 and 80.000, respectively)

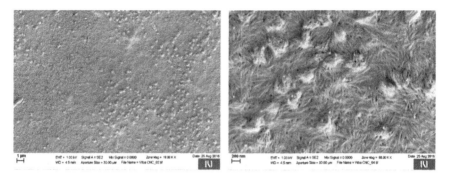

Fig. 6.13 SEM image of CNC extracted from weeping willow (*Salix babylonica*), (the magnification 10.000 and 60.000, respectively)

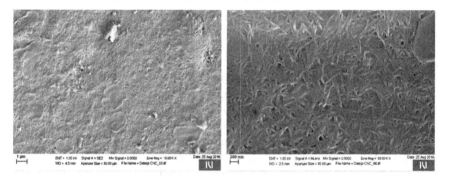

Fig. 6.14 SEM image of CNC extracted from lucerne (*Medicago sativa*), (the magnification 15.000 and 60.000, respectively)

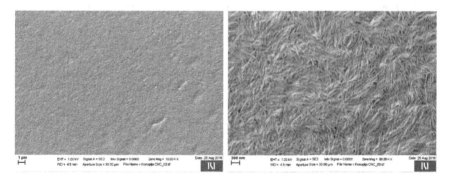

Fig. 6.15 SEM image of CNC extracted from hemp (*Cannabis sativa*), (the magnification 10.000 and 60.000, respectively)

6.3.2 Properties of Nano-Crystalline Cellulose

The particle size and size distribution analyses were performed using a Zetasizer Nanoseries particle size analyzer (Malvern Instruments Ltd., Malvern, UK). The measuring conditions were: T = 25 °C, measuring angle 173° and the wave length 633 nm. The mean particle diameter of each sample was reported as the average value of three replicates. The results are collected in Table 6.3.

Although the design of nanoparticles is rod-loke as observed on SEM photographs, the analytical method employed for determination of particles sizes, is based on the assumption of spherical particles, therefore the obtained results are thus only relative values which serve to compare the samples with each other.

The smallest particles were obtained from lucerne, with the average size of 205 nm, weeping willow with slightly larger particles (the average size is 270 nm) followed. Dimensions of nano-cellulose extracted from Spanish broom (average size is 338 nm), and from hemp (average size is 397 nm) were in the range between 330 nm and 400 nm, however the largest particles were observed for balm-leaved archangel and nettle (average size is 442 and 569 nm, respectively). Particles are not homogeneous in their sizes, which is confirmed by high variation coefficients between 40 and 60%. The results are in good agreement with the research of nano-cellulose carried out by other researchers. Chang with co-authors prepared nano-cellulose by different conditions of acid hydrolysis from cotton linter. Sizes of nanocrystals obtained by the hydrolysis with 50 and 55% sulfuric acid determined by dynamic light scattering were between 91–295 nm and 164–615 nm, respectively [10]. Espino et al. prepared CNC from fibres extracted from agave T and barley husks, particles were analysed and compared to particles obtained from microcrystalline cellulose MCC. Monodispersed particles were obtained and their dimensions were determined by dynamic light scattering. Dimensions of CNC obtained from MCC, agave T and barley were 188 ± 13 nm, 214 ± 19 nm, 269 ± 16 nm, respectively with Polydispersity index (PDI) 0.22, 0.21 and 0.11

Table 6.3 Nano-particles' sizes of nettle (*Urtica dioica*), Spanish broom (*Spartium junceum*), balm-leaved archangel (*Lamium orvala*) and lucerne (*Medicago sativa*), weeping willow (*Salix babylonica*) and hemp (*Cannabis sativa*)

Fibre type	Average particle size (nm)	Standard deviation (nm)	PDI	Zeta potential mV
Nettle	569	200	0.5	−30.8
Spanish broom	338	149	0.5	−37.5
Balm-leaved archangel	442	216	0.6	−38.8
Lucerne	205	85	0.6	−39.4
Weeping willow	270	22	0.3	−34.9
Hemp	397	254	0.4	−23.0

respectively. Sizes of CNC from agricultural by-products are larger than particles obtained from MCC [11].

Polydispersity index PDI is a measure of the total particle size distribution. Values of PDI < 0.1 are indicative of good homogeneity of the sample and of a narrow size distribution of nanoparticles, however PDI values >0.3 represent samples heterogeneity [10]. PDI of nanoparticles' samples extracted from nettle, lucerne, balm-leaved archangel, weeping willov, Spanish broom and hemp was about 0.5, which is expected for natural plant fibres. The broad size distribution is arising from structural variations of the starting materials with nonuniform crystalline morphology. From the organic structurally diverse materials it is very difficult to obtain uniform-sized particles.

Very important for nanoparticles preparation is if they aggregate or not. According to the DLVO (Derjaguin, Landau, Verwey and Overbeek) theory, which represents an important framework to model interactions in aqueous colloidal suspensions and the respective aggregation rates, colloid stability depends on the sum of van der Waals attractive forces and electrostatic repulsive forces [12]. Zeta potential ZP, which is the potential difference between the phase boundaries of solid substances and liquids, provides information on the electrostatic repulsive forces. As it is a measure of the electric charge of the particles, which are suspended in the liquid, it is commonly related with colloid stability. Nano-particles dispersions with ZP values of ±0–10 mV, ±10–20 mV and ±20–30 mV and >± 30 mV are commonly classified as highly unstable, relatively stable, moderately stable and highly stable, respectively [13]. Although ZP does not provide any insight on the attractive van der Waals forces, we concluded based on the obtained ZP values for dispersions of nettle, lucerne, weeping willow, Spanish broom and balm-leaved archangel (ZP = −30 mV to −40 mV) that highly stable dispersions of nanoparticles from various cellulosic sources were produced. ZP of hemp nano-cellulose dispersersion, that is in the range ZP = −20 mV to −30 mV, is indicating on a moderate stability of hemp nano-cellulose particles.

References

1. Östbom G, Svennerstedt B (2007) Fiber quality of hemp grown on the Swedish Island Gotland. In: Anandjiwala R, Hunter L (ed) Textiles for sustainable development. Nova Science Publishers Inc. pp 53–61
2. Sankari HS (2000) Comparison of bast fibre yield and mechanical fibre properties of hemp (*Cannabis sativa* L.) cultivars. Ind Crops Prod 11:73–84
3. Kostic M, Pejic B, Skundric P (2008) Quality of chemically modified hemp fibers. Biores Technol 99:94–99
4. Tadele A (2017) Effect of maturity on fundamental properties of nettle fibres. J Text Fashion Technol 3(1):1–10
5. Kovacevic Z, Bischof S, Vujasinovic E, Fan M (2016) The influence of pre-treatment of *Spartium junceum* L. fibres on the structure and mechanical properties of biocomposites. https://doi.org/10.1016/j.arabjc.2016.08.004

6. Cerchiara T, Chidichimo G, Rondi G et al (2014) Chemical composition, morphology and tensile properties of Spanish Broom (*Spartium junceum* L.) fibres in comparison with flax (*Linum usitatissimum* L.). Fibres Text Eastern Europe 2(22): 25–28

7. Segal LJJ, Martin AE, Conrad CM (1959) An empirical method for estimating the degree of crystallinity of native cellulose using the X-ray diffractometer. Text Res J 9(10)

8. Park S, Baker JO, Himmel ME et al (2010) Cellulose crystallinity index: measurement techniques and their impact on interpreting cellulase performance. Biotechnol Biofuels 3(10). https://doi.org/10.1186/1754-6834-3-10

9. Terinte N, Ibbett R, Schuster KC (2011) Overview on native cellulose and microcrystalline cellulose 1 structure studied by X-ray diffraction (WAXD): comparison between measurements techniques. Lenzinger Berichte 89:118–131

10. Chang C, Wang I, Hung K et al (2010) Preparation and characterization of nanocrystalline cellulose by acid hydrolysis of cotton linter. Taiwan J For Sci 25(3):231–244

11. Espino E, Cakir M, Domenek S et al (2014) Isolation and characterization of cellulose nanocrystals from industrial by-products of *Agave tequilana* and barley. Ind Crops Prod 62:552–559

12. Trefalt G, Borkovec M (2014) Overview of DLVO Theory. www.colloid.ch/dlvo. Accessed 20 Oct 2016

13. Bhattacharjee S (2016) DLS and zeta potential—what they are and what they are not? J Controlled Release 235:337–351

Chapter 7
Conclusion

Abstract Sustainable materials based on renewable raw materials with new or improved properties are currently the research imperative, however the exploitation of non-traditional, especially wild plants for fibre extraction or for the production of cellulose may be associated with a number of concerns.

Keywords Sustainable fibres · Concerns

Sustainable materials based on renewable raw materials with new or improved properties are currently the research imperative. Different plants represent in this respect an ideal base in the field of fibrous materials. Traditionally represent the group of fibre yielding plants cotton, flax, hemp, jute, coir, sisal and a few more, however there are still many other plants with the potential for fibres extraction but they have not yet found their place in this group. Recently other non-conventional plants are studied with the aim to extract and characterize fibres to evaluate their applicability. So less common, but promising materials are developed mainly as cellulose based composites or nanocomposites, which is one of the most promising research fields related to plant-based products.

The world community is looking at cost-efficient ways to make the economies more climate-friendly and less energy-consuming (EU Climate Strategies and Targets), however, there are many obstacles that need to be taken into account considering the specific nature of certain cases. The exploitation of non-traditional, especially wild plants for fibre extraction or for the production of cellulose may be associated with a number of concerns:

(i) Cost-efficiency is conditioned by the costs of harvesting and costs of fibre extraction, especially where quantities are smaller and fibre extraction is done manually in underdeveloped economies. In addition, transport costs are not negligible.

(ii) The massive exploitation of wild plants may also have an impact on biodiversity and ecosystem changes if the maintenance of the balance between the exploitation and conservation of plant crops is insufficient.

M. Sfiligoj Smole et al., *Surface Properties of Non-conventional*
Cellulose Fibres, Biobased Polymers, https://doi.org/10.1007/978-3-030-10407-8_7

(iii) The use of agro-waste positively influences the removal of these, but the procedures of cellulose extraction are usually related to the use of water and various chemicals, which again represents the environmental impact.

(iv) Plant yields are variable if the growth of wild plants is uncontrolled. The yield may be highly changeable depending on the annual weather influences, etc. In addition to, cultivation of wild plants on agricultural land is otherwise possible when it proves to be profitable.

(v) Quality assurance of raw materials, e.g. lengths of fibers, fineness, length distribution, maturity of fibers, chemical composition, purity of raw materials etc. is also questionable in wild growing plants and agro-waste.

During natural sources utilization, one should always take into account environment impacts. Life-cycle assessment (LCA) is a technique to assess environmental impacts associated with all the stages of a product's life from raw material extraction through materials processing, manufacture, distribution, use, repair and maintenance, and disposal or recycling.

LCA studies were performed on agro wastes, e.g. hemp hurds for bio-ethanol production (e.g. [1–4]) and on cellulose composites and nanocomposites (e.g. [5]).

Life cycle assessment (LCA) studies of polysaccharide materials for textiles, engineering materials and packing in order to gain insight of the environmental profiles of these products (e.g. viscose or natural fibre polymer composites) in comparison with their conventional counterparts (e.g. cotton or petrochemical polymers) were reviewed by [6]. The results show advantages of natural and man-made cellulose fibres when compared to synthetics. Authors conclude that from cradle to grave, in terms of non-renewable energy requirements and greenhouse gas emissions, the polysaccharide products are better than their conventional counterparts, which are mostly petrochemical-based materials. Cotton is the exception because its cultivation requires a relatively high amount of water and chemicals 6.

References

1. González-García S, Luo L, Moreira MT et al (2012) Life cycle assessment of hemp hurds use in second generation ethanol production. Biomass Bioenergy 36:268–279
2. Restrepo A, Becerra R, Tibaquirá JEG (2016) Energetic and carbon footprint analysis in manufacturing process of bamboo boards in Colombia
3. Agyekum EO, Fortuin KPJ, van der Harst E (2017) Environmental and social life cycle assessment of bamboo bicycle frames made in Ghana. J Cleaner Prod 143:1069–1080. https://doi.org/10.1016/j.jclepro.2016.12.012
4. Vogtländer J, van der Lugt P, Brezet H (2010) The sustainability of bamboo products for local and Western European applications. LCAs and land-use. J Clean Prod 18:1260–1269
5. Arvidsson R, Nguyen D, Svanström M (2015) Life cycle assessment of cellulose nanofibrils production by mechanical treatment and two different pretreatment processes. Environ Sci Technol 49(11):6881–6890. https://doi.org/10.1021/acs.est.5b00888
6. Shen L, Patel MK (2008) Life cycle assessment of polysaccharide materials: A review. J Polym Environ 16(2):154-167. https://doi.org/10.1007/s10924-008-0092-9

Printed in the United States
By Bookmasters